T0406791

Advanced Structured Materials

Volume 43

Series Editors

Andreas Öchsner
Lucas F. M. da Silva
Holm Altenbach

For further volumes:
http://www.springer.com/series/8611

Pablo Andrés Muñoz-Rojas
Editor

Optimization of Structures and Components

Editor
Pablo Andrés Muñoz-Rojas
Department of Mechanical Engineering
Santa Catarina State University—(UDESC)
Campus Universitário Avelino Marcante
Joinville, Santa Catarina
Brazil

ISSN 1869-8433 ISSN 1869-8441 (electronic)
ISBN 978-3-319-00716-8 ISBN 978-3-319-00717-5 (eBook)
DOI 10.1007/978-3-319-00717-5
Springer Cham Heidelberg New York Dordrecht London

Library of Congress Control Number: 2013945292

Printed on acid-free paper

Springer is part of Springer Science+Business Media (www.springer.com)

Preface

The term *Structural Optimization* was coined by Lucien Schmit in 1960, when he published an innovative paper[1] and launched a revolution in structural design procedures. His idea consisted in coupling nonlinear numerical optimization techniques, typically used in operations research, to finite element structural analyses, in order to achieve an optimal design. Although the origin of optimization methods traces back to the days of Newton, Leibnitz, and Cauchy, it was Schmit's contribution that settled a mark for the development and application of modern optimization techniques in engineering. Certainly, the parallel evolution of computer technology and its rapid dissemination in the following decades were deterministic for the success of this new design strategy.

One example of structural optimization has to do with minimization of weight, which has always been a major concern in aeronautical and aerospace engineering. More recently, the automotive industry has also focused in developing lighter vehicles, since engine technology has reached such a maturity level, that efficiency in fuel consumption is now guided by weight reduction. Following this trend, in modern world, competition for a market share requires better products with higher efficiency and lower costs, so that new design challenges are set every day. In this context, it is important that both, researchers and practitioning engineers, be continuously developing and applying optimization methods to solve problems that are ever increasing in complexity and computer storage needs. In spite of all the evolution that this area has experienced, several design problems still cannot be solved in an efficient manner.

In order to have a good structural optimization method, quite a large number of factors must be accounted for: the selection of an overall strategy (size, shape or topology optimization); the selection of an appropriate objective function, constraints and design variables; the selection of an approximation technique for the functions involved and the selection of a robust optimization algorithm. If the optimization algorithm requires derivatives, an efficient and accurate sensitivity analysis procedure is also necessary.

---

[1] Schmit, L. A.: Structural Design by Systematic Synthesis, Proceedings, 2nd Conference on Electronic Computation, ASCE, New York, pp. 105–122, 1960.

In this book, eight chapters discuss recent and sometimes original developments in one of the aforementioned aspects, ranging from analytical expressions for topological derivatives to the application of meta-models for the solution of large scale problems. The authors were asked to decide about the length of their chapters, such that the presentation of the kernel ideas would not be sacrificed by space limitations. Hence, it is expected that the contents can be adopted as referential texts due to their detail level.

I would like to thank sincerely the commitment of all the contributors who made this text possible. Their patient cooperation at all stages of the book project until the final form of the manuscripts was crucial to the high quality of the result.

April 2013 Pablo Andrés Muñoz-Rojas

Contents

Topological Derivative for Multi-Scale Linear Elasticity Models in Three Spatial Dimensions

Antonio André Novotny

Abstract A remarkably simple analytical expression for the sensitivity of the three-dimensional macroscopic elasticity tensor to topological microstructural changes of the underlying material is obtained. The derivation of the proposed formula relies on the concept of topological derivative, applied within a variational multi-scale constitutive framework where the macroscopic strain and stress at each point of the macroscopic continuum are volume averages of their microscopic counterparts over a representative volume element (RVE) of material associated with that point. The derived sensitivity, given by a symmetric fourth order tensor field over the microstructure domain, measures how the estimated three-dimensional macroscopic elasticity tensor changes when a small spherical void is introduced at the micro-scale level. The obtained result can be applied in the synthesis and optimal design of three-dimensional elastic microstructures.

1 Introduction

The accurate prediction of the *constitutive behavior* of a continuum body under loading is of paramount importance in many areas of engineering and science. Until about a decade ago, this issue has been addressed mainly by means of conventional *phenomenological constitutive theories*. More recently, the increasing understanding of the microscopic mechanisms responsible for the macroscopic response, allied to the industrial demand for more accurate predictive tools, led to the development and use of so-called *multiscale constitutive theories*. Such theories are currently a subject of intensive research in applied mathematics and computational mechanics. Their starting point can be traced back to the pioneering developments reported in

A. A. Novotny (✉)
Laboratório Nacional de Computação Científica LNCC/MCT, Coordenação de Matemática Aplicada e Computacional, Av. Getúlio Vargas 333, Petrópolis - RJ 25651-075, Brasil
e-mail: novotny@lncc.br

P. A. Muñoz-Rojas (ed.), *Optimization of Structures and Components*,
Advanced Structured Materials 43, DOI: 10.1007/978-3-319-00717-5_1,

[1–6]. Early applications were concerned with the description of relatively simple microscale phenomena often treated by analytical or semi-analytical methods [7–12]. More recent applications rely often on finite element-based computational simulations [13, 14] and are frequently applied to more *complex material behavior* in areas such as the modeling of human arterial tissue [15], bone [16], the plastic behavior of porous metals [17] and the microstructural evolution and phase transition in the solidification of metals [18]. The reader may refer to [19] for an introduction to the multiscale constitutive modeling from the point of view of solid mechanics, and to [5] for the mathematical description of such modeling.

One interesting branching of such developments is the study of the sensitivity of the macroscopic response to changes in the underlying microstructure. The sensitivity information becomes essential in the analysis and potential purpose-design and optimization of heterogenous media. For instance, sensitivity information obtained by means of a relaxation-based technique has been successfully used in [20–22] to design microstructural topologies with negative macroscopic Poisson's ratio. Multiscale models have also been applied with success to the topology optimization of load bearing structures in the context of the so-called homogenization approach to topology optimization (see, for instance, the review paper by Eschenauer and Olhoff [23] and the book by Allaire [24]) based on the fundamental papers by Żochowski [25] and Bendsøe and Kikuchi [26]. See also [27, 28]. In such cases, the microscale model acts as a regularization of the exact problem posed by a material point turning into a hole [29]. The homogenization approach has also been applied to microstructural topology optimization problems where the target is the design of topologies that yield pre-specified or extreme macroscopic response [30–32]. One of the drawbacks of this methodology, however, is that it often produces designs with large regions consisting of perforated material. To deal with this problem, a penalization of intermediate densities is commonly introduced.

In contrast to the regularized approaches, in this paper we present a general *exact* analytical expression for the sensitivity of the three-dimensional macroscopic elasticity tensor to topological changes of the microstructure of the underlying material. We follow the ideas developed in [33] for two-dimensional elasticity. The obtained sensitivity is given by a symmetric fourth order tensor field over the representative volume element (RVE) that measures how the macroscopic elasticity constants estimated within the multiscale framework changes when a small spherical void is introduced at the microscale. It is derived by making use of the notions of topological asymptotic analysis and topological derivative [34] within the variational formulation of well-established multiscale constitutive theory fully developed in the book by Sanchez-Palencia [5] (see also [13, 14, 35]), where the macroscopic strain and stress tensors are volume averages of their microscopic counterparts over a RVE of material. The final format of the proposed analytical formula is strikingly simple and can be used in applications such as the synthesis and optimal design of microstructures to meet a specified macroscopic behavior [36].

2 The Homogenized Elasticity Tensor

The *homogenization-based multiscale constitutive framework* presented, among others, in [5], is adopted here in the estimation of the *macrostructure* elastic response from the knowledge of the underlying *microstructure*. The main idea behind this well-established family of constitutive theories is the assumption that any point x of the macroscopic continuum is associated to a local *RVE* whose domain is denoted by Ω_μ, with boundary $\partial\Omega_\mu$, as shown in Fig. 1. Crucial to the developments presented in this paper is the closed form of the *homogenized elasticity tensor* $\mathbb{C}$ for the multiscale model defined in the above. The components of the homogenized elasticity tensor $\mathbb{C}$, in the orthonormal basis $\{e_i\}$, for $i = 1, 2, 3$, of the Euclidean space (refer to Fig. 1), can be written as

$$(\mathbb{C})_{ijkl} = \frac{1}{V_\mu} \int_{\Omega_\mu} (\sigma_\mu(u_{\mu_{kl}}))_{ij}, \tag{1}$$

where V_μ denotes the total volume of the RVE, e.g. the volume of the cub in the sketch shown in Fig. 1. The canonical microscopic displacement field $u_{\mu_{kl}}$ is solution to the equilibrium equation of the form [5]

$$\int_{\Omega_\mu} \sigma_\mu(u_{\mu_{kl}}) \cdot \nabla\eta^s = 0 \qquad \forall \eta \in \mathcal{V}_\mu. \tag{2}$$

We assume that the microscopic stress tensor field $\sigma_\mu(u_{\mu_{kl}})$ satisfies

$$\sigma_\mu(u_{\mu_{kl}}) = \mathbb{C}_\mu \nabla u^s_{\mu_{kl}}, \tag{3}$$

where $\mathbb{C}_\mu$ is the microscopic constitutive tensor given by

$$\mathbb{C}_\mu = \frac{E_\mu}{1+\nu_\mu}\left(\mathbb{I} + \frac{\nu_\mu}{1-2\nu_\mu} \mathrm{I} \otimes \mathrm{I}\right), \tag{4}$$

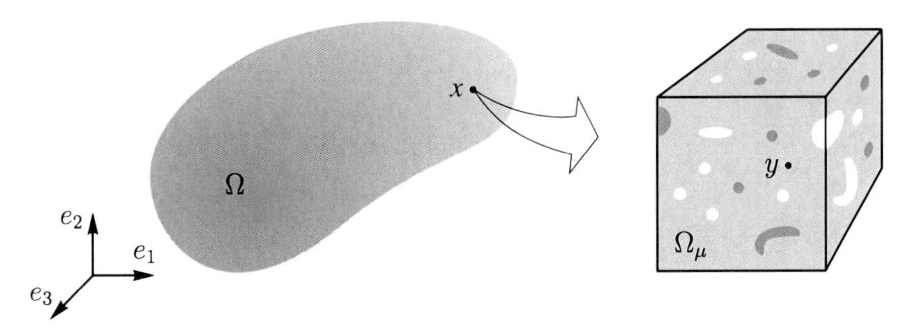

Fig. 1 Macroscopic continuum with a locally attached microstructure

with I and $\mathbb{I}$ the second and fourth order identity tensors, respectively, E_μ the Young modulus and ν_μ Poisson ratio of the RVE.

Without loss of generality, $u_{\mu_{kl}}(y)$, with $y \in \Omega_\mu$, may be decomposed into a sum

$$u_{\mu_{kl}}(y) := u + (e_k \otimes e_l)y + \widetilde{u}_{\mu_{kl}}(y), \tag{5}$$

of a constant (rigid) RVE displacement coinciding with the macroscopic displacement field u at the point $x \in \Omega$, a linear field $(e_k \otimes e_l)y$, and a canonical microscopic displacement fluctuation field $\widetilde{u}_{\mu_{kl}}(y)$. The microscopic displacement fluctuation field $\widetilde{u}_{\mu_{kl}}$ is solution to the following *canonical set of variational problems* [5]:

$$\left\{\begin{array}{l} \text{Find } \widetilde{u}_{\mu_{kl}} \in \mathscr{V}_\mu, \text{ such that} \\ \displaystyle\int_{\Omega_\mu} \sigma_\mu(\widetilde{u}_{\mu_{kl}}) \cdot \nabla\eta^s + \int_{\Omega_\mu} \mathbb{C}_\mu(e_k \otimes_s e_l) \cdot \nabla\eta^s = 0 \quad \forall \eta \in \mathscr{V}_\mu, \\ \text{with } \ \sigma_\mu(\widetilde{u}_{\mu_{kl}}) = \mathbb{C}_\mu \nabla\widetilde{u}_{\mu_{kl}}. \end{array}\right. \tag{6}$$

The complete characterization of the multiscale constitutive model is obtained by defining the subspace $\mathscr{V}_\mu \subset \mathscr{U}_\mu$ of kinematically admissible displacement fluctuations. In general, different choices produce different macroscopic responses for the same RVE. In this section, the analysis will be focussed on media with periodic microstructure. In this case, the geometry of the RVE cannot be arbitrary and must represent a cell whose periodic repetition generates the macroscopic continuum. In addition, the displacement fluctuations must satisfy periodicity on the boundary of the RVE. Accordingly, we have

$$\mathscr{V}_\mu := \left\{\varphi \in \mathscr{U}_\mu : \varphi(y^+) = \varphi(y^-) \quad \forall\, (y^+, y^-) \in \mathfrak{P}\right\}, \tag{7}$$

where $\mathfrak{P}$ is the set of pairs of points, defined by a one-to-one periodicity correspondence, lying on opposing sides of the RVE boundary. Finally, the *minimally constrained space* of kinematically admissible displacements $\mathscr{U}_\mu$ is defined as

$$\mathscr{U}_\mu := \left\{\varphi \in H^1(\Omega_\mu; \mathbb{R}^3) : \int_{\Omega_\mu} \varphi = 0, \ \int_{\partial\Omega_\mu} \varphi \otimes_s n = 0\right\}. \tag{8}$$

where n is the outward unit normal to the boundary $\partial\Omega_\mu$ and $\otimes_s$ denotes the symmetric tensor product between vectors.

From the definition of the homogenized elasticity tensor (1), we have

$$\begin{aligned} (\mathbb{C})_{ijkl} &= \frac{1}{V_\mu}\int_{\Omega_\mu} e_i \cdot \sigma_\mu(u_{\mu_{kl}})e_j \\ &= \frac{1}{V_\mu}\int_{\Omega_\mu} \sigma_\mu(u_{\mu_{kl}}) \cdot (e_i \otimes e_j). \end{aligned} \tag{9}$$

On the other hand, the additive decomposition (5) allows us to write

$$\begin{aligned} e_i \otimes e_j &= \nabla((e_i \otimes e_j)y)^s \\ &= \nabla(u_{\mu_{ij}}(y) - \widetilde{u}_{\mu_{ij}}(y) - u)^s \\ &= \nabla(u_{\mu_{ij}}(y) - \widetilde{u}_{\mu_{ij}}(y))^s . \end{aligned} \quad (10)$$

since $(e_i \otimes e_j)y = u_{\mu_{ij}}(y) - \widetilde{u}_{\mu_{ij}}(y) - u$. Therefore, by combining these two results, we obtain

$$\begin{aligned} (\mathbb{C})_{ijkl} &= \frac{1}{V_\mu} \int_{\Omega_\mu} \sigma_\mu(u_{\mu_{kl}}) \cdot \nabla(u_{\mu_{ij}} - \widetilde{u}_{\mu_{ij}})^s \\ &= \frac{1}{V_\mu} \int_{\Omega_\mu} \sigma_\mu(u_{\mu_{kl}}) \cdot \nabla u^s_{\mu_{ij}}, \end{aligned} \quad (11)$$

since $u_{\mu_{kl}}$ satisfies the equilibrium equation (2) and $\widetilde{u}_{\mu_{ij}} \in \mathscr{V}_\mu$.

3 Sensitivity of the Macroscopic Elasticity Tensor to Topological Microstructural Changes

A closed formula for the sensitivity of the homogenized elasticity tensor (1) to the nucleation of a spherical cavity within the RVE is presented in this section. We start by noting that each component of the homogenized elasticity tensor is defined by the energy based functional (11). Therefore, by taking into account the result derived in [37], the topological derivative of (11) with respect to the nucleation of a spherical cavity at an arbitrary point $\widehat{y} \in \Omega_\mu$ is given by

$$\mathscr{T}_\mu(\widehat{y}) = -\mathbb{P}_\mu \sigma_\mu(u_{\mu_{ij}}(\widehat{y})) \cdot \sigma_\mu(u_{\mu_{kl}}(\widehat{y})), \quad \forall \widehat{y} \in \Omega_\mu, \quad (12)$$

with the *polarization tensor* $\mathbb{P}_\mu$ given by

$$\mathbb{P}_\mu = \frac{3}{2V_\mu E_\mu} \left(10 \frac{1-\nu_\mu^2}{7-5\nu_\mu} \mathbb{I} - \frac{(1-\nu_\mu)(1+5\nu_\mu)}{7-5\nu_\mu} \mathrm{I} \otimes \mathrm{I} \right). \quad (13)$$

From the result given by (12), we can recognize a fourth order tensor field over Ω_μ that represents the sensitivity of the macroscopic elasticity tensor $\mathbb{C}$ to topological microstructural changes resulting from the insertion of a spherical cavity within the RVE. Therefore, the *topological derivative* of the homogenized elasticity tensor reads

$$\mathbb{T}_\mu(\widehat{y}) = -\mathbb{P}_\mu \sigma_\mu(u_{\mu_{ij}}(\widehat{y})) \cdot \sigma_\mu(u_{\mu_{kl}}(\widehat{y})) e_i \otimes e_j \otimes e_k \otimes e_l, \quad (14)$$

where the fields $u_{\mu_{ij}}$ come out from the solutions to (6) for the unperturbed RVE domain Ω_μ together with the additive decomposition (5).

Remark 1 The remarkable simplicity of the closed form sensitivity given by (14) is to be noted. Once the vector fields $\widetilde{u}_{\mu_{ij}}$ have been obtained as solutions to the set of variational equations (6) for the *original* RVE domain, the sensitivity tensor $\mathbb{T}_\mu$ can be trivially assembled from the additive decomposition (5). The information provided by the fourth order topological derivative tensor field $\mathbb{T}_\mu$ given by (14) can be used in a number of practical applications such as the design of microstructures to match a specified macroscopic constitutive response.

4 Conclusions

Expression (14) allows the *exact* topological derivative of any differentiable function of $\mathbb{C}$ be calculated through the direct application of the conventional rules of differential calculus. That is, any such a function $\Psi(\mathbb{C})$ has exact topological derivative of the form

$$\mathscr{T}_\mu = \langle D\Psi(\mathbb{C}), \mathbb{T}_\mu \rangle, \tag{15}$$

with the brackets $\langle \cdot, \cdot \rangle$ denoting the appropriate product between the derivative of Ψ with respect to $\mathbb{C}$ and the topological derivative $\mathbb{T}_\mu$ of $\mathbb{C}$. Note, for example, that properties of interest such as the homogenized Young's, shear and bulk moduli as well as the Poisson ratio are all regular functions of $\mathbb{C}$. This observation together with Remark 1 point strongly to the suitability of the use of (15) in a topology algorithm for the synthesis and optimization of elastic micro-structures based on the minimization/maximization of cost functions defined in terms of homogenized properties. These features has been successfully explored in [36] in two-dimensional multiscale elasticity problem.

In order to fix theses ideas, let us present three examples concerning the topological derivatives of given functions $\Psi(\mathbb{C})$. Let $\varphi_1, \varphi_2 \in \mathbb{R}^3 \times \mathbb{R}^3$ be any pair of second order tensors. Then we obtain the following results, which can be used in numerical methods of synthesis and/or topology design of microstructures [36]:

Example 1 We consider a function $\Psi(\mathbb{C})$ of the form

$$\Psi(\mathbb{C}) := \mathbb{C}\varphi_1 \cdot \varphi_2, \tag{16}$$

Therefore, according to (15), its topological derivative is given by

$$\mathscr{T}_\mu = \mathbb{T}_\mu \varphi_1 \cdot \varphi_2. \tag{17}$$

If we set $\varphi_1 = e_i \otimes e_j$ and $\varphi_2 = e_k \otimes e_l$, for instance, we get $\Psi(\mathbb{C}) = (\mathbb{C})_{ijkl}$ and its topological derivative is given by $\mathscr{T}_\mu = (\mathbb{T}_\mu)_{ijkl}$. It means that $\mathscr{T}_\mu$ actually represents the topological derivative of the component $(\mathbb{C})_{ijkl}$ of the homogenized elasticity tensor $\mathbb{C}$.

Example 2 Now, let us consider a function $\Psi(\mathbb{C})$ of the form

$$\Psi(\mathbb{C}) := \mathbb{C}^{-1}\varphi_1 \cdot \varphi_2. \tag{18}$$

According again to (15), the topological derivative of $\Psi(\mathbb{C})$ is given by

$$\mathscr{T}_\mu = -(\mathbb{C}^{-1}\mathbb{T}_\mu\mathbb{C}^{-1})\varphi_1 \cdot \varphi_2. \tag{19}$$

Thus, by setting tensors φ_1 and φ_2 properly, we can obtain the topological derivative in its explicit form of any component of the inverse of the homogenized elasticity tensor $\mathbb{C}^{-1}$. The above derivation requires some additional explanation. Note that we can differentiate the relation $\mathbb{C}\mathbb{C}^{-1} = \mathbb{I}$ with respect to $\mathbb{C}$, namely

$$\mathbb{T}_\mu\mathbb{C}^{-1} + \mathbb{C}D(\mathbb{C}^{-1}) = 0. \tag{20}$$

After multiplying to the left by $\mathbb{C}^{-1}$ we get

$$\mathbb{C}^{-1}\mathbb{T}_\mu\mathbb{C}^{-1} + D(\mathbb{C}^{-1}) = 0, \tag{21}$$

which leads to

$$D(\mathbb{C}^{-1}) = -\mathbb{C}^{-1}\mathbb{T}_\mu\mathbb{C}^{-1}. \tag{22}$$

Example 3 Finally, we consider a function $\Psi(\mathbb{C})$ of the form

$$\Psi(\mathbb{C}) := \frac{\mathbb{C}^{-1}\varphi_1 \cdot \varphi_2}{\mathbb{C}^{-1}\varphi_1 \cdot \varphi_1} + \frac{\mathbb{C}^{-1}\varphi_2 \cdot \varphi_1}{\mathbb{C}^{-1}\varphi_2 \cdot \varphi_2}. \tag{23}$$

From 15, the corresponding topological derivative is

$$\begin{aligned} \mathcal{T}_\mu = &- \frac{(\mathbb{C}^{-1}\mathbb{T}_\mu\mathbb{C}^{-1})\varphi_1 \cdot [(\mathbb{C}^{-1}\varphi_1 \cdot \varphi_1)\varphi_2 - (\mathbb{C}^{-1}\varphi_1 \cdot \varphi_2)\varphi_1]}{(\mathbb{C}^{-1}\varphi_1 \cdot \varphi_1)^2} \\ &- \frac{(\mathbb{C}^{-1}\mathbb{T}_\mu\mathbb{C}^{-1})\varphi_2 \cdot [(\mathbb{C}^{-1}\varphi_2 \cdot \varphi_2)\varphi_1 - (\mathbb{C}^{-1}\varphi_2 \cdot \varphi_1)\varphi_2]}{(\mathbb{C}^{-1}\varphi_2 \cdot \varphi_2)^2}. \end{aligned} \tag{24}$$

References

1. Bensoussan, A., Lions, J.L., Papanicolau, G.: Asymptotic Analysis for Periodic Microstructures. Elsevier, Amsterdam (1978)
2. Hashin, Z., Shtrikman, S.: A variational approach to the theory of the elastic behaviour of multiphase materials. J. Mech. Phys. Solids **11**(2), 127–140 (1963)
3. Hill, R.: A self-consistent mechanics of composite materials. J. Mech. Phys. Solids **13**(4), 213–222 (1965)
4. Mandel, J.: Plasticité classique et viscoplasticité. Springer, Udine (1971). (CISM Lecture Notes)

5. Sanchez-Palencia, E.: Non-homogeneous Media and Vibration Theory, volume 127 of Lecture Notes in Physics. Springer, Berlin (1980)
6. Suquet, P.M.: Elements of Homogenization for Inelastic Solid Mechanics, volume 272 of Homogenization Techniques for Composite Media, Lecture Notes in Physics 272. Springer, Berlin (1987)
7. Auriault, J.L.: Effective macroscopic description for heat conduction in periodic composites. Int. J. Heat Mass Transfer **26**(6), 861–869 (1983)
8. Auriault, J.L., Royer, P.: Double conductivity media: a comparison between phenomenological and homogenization approaches. Int. J. Heat Mass Transfer **36**(10), 2613–2621 (1993)
9. Gurson, A.L.: Continuum theory of ductile rupture by void nucleation and growth: Part I—yield criteria and flow rule for porous ductile media. J. Eng. Mater. Technol. Trans. ASME **99**(1), 2–15 (1977)
10. Nemat-Nasser, S.: Averaging theorems in finite deformation plasticity. Mech. Mater. **31**(8), 493–523 (1999)
11. Nemat-Nasser, S., Hori, M.: Micromechanics: Overall Properties of Heterogeneous Materials. Elsevier, Amsterdam (1993)
12. Ostoja-Starzewski, M., Schulte, J.: Bounding of effective thermal conductivities of multiscale materials by essential and natural boundary conditions. Phys. Rev. B **54**(1), 278–285 (1996)
13. Michel, J.C., Moulinec, H., Suquet, P.: Effective properties of composite materials with periodic microstructure: a computational approach. Comput. Meth. Appl. Mech. Eng. **172**(1–4), 109–143 (1999)
14. Miehe, C., Schotte, J., Schröder, J.: Computational micro-macro transitions and overall moduli in the analysis of polycrystals at large strains. Comput. Mater. Sci. **16**(1–4), 372–382 (1999)
15. Speirs, D.C.D., de Souza Neto, E.A., Perić, D.: An approach to the mechanical constitutive modelling of arterial tissue based on homogenization and optimization. J. Biomech. **41**(12), 2673–2680 (2008)
16. Oyen, M.L., Ferguson, V.L., Bembey, A.K., Bushby, A.J., Boyde, A.: Composite bounds on the elastic modulus of bone. J. Biomech. **41**(11), 2585–2588 (2008)
17. Giusti, S.M., Blanco, P.J., de Souza Neto, E.A., Feijóo, R.A.: An assessment of the Gurson yield criterion by a computational multi-scale approach. Eng. Comput. **26**(3), 281–301 (2009)
18. Celentano, D.J., Dardati, P.M., Godoy, L.A., Boeri, R.E.: Computational simulation of microstructure evolution during solidification of ductile cast iron. Int. J. Cast Met. Res. **21**(6), 416–426 (2008)
19. Lewinski, T., Telega, J.J.: Plates, laminates, and shells: asymptotic analysis and homogenization. World Scientific (2000)
20. Almgreen, R.F.: An isotropic three-dimensional structure with Poisson's ratio—1. J. Elast. **15**(4), 427–430 (1985)
21. Lakes, R.: Foam structures with negative Poisson's ratio. Science AAAS **235**(4792), 1038–1040 (1987)
22. Lakes, R.: Negative Poisson's ratio materials. Science AAAS **238**(4826), 551 (1987)
23. Eschenauer, H.A., Olhoff, N.: Topology optmization of continuum structures: a review. Appl. Mech. Rev. **54**(4), 331–390 (2001)
24. Allaire, G.: Shape Optimization by the Homogenization Method, volume 146 of Applied Mathematical Sciences. Springer, New York (2002)
25. Żochowski, A.: Optimal perforation design in 2-dimensional elasticity. Mech. Struct. Mach. **16**(1), 17–33 (1988)
26. Bendsøe, M.P., Kikuchi, N.: Generating optimal topologies in structural design using an homogenization method. Comput. Meth. Appl. Mech. Eng. **71**(2), 197–224 (1988)
27. Czarnecki, S., Lewinski, T.: A stress-based formulation of the free material design problem with the trace constraint and single loading condition. Bull. Pol. Acad. Sci. Tech. Sci. 60(2), (2012)
28. Nowak, M.: Structural optimization system based on trabecular bone surface adaptation. Struct. Multi Optim. **32**(3), 241–249 (2006)

29. Belytschko, T., Xiao, S., Parimi, C.: Topology optimization with implicit functions and regularization. Int. J. Numer. Meth. Eng. **57**, 1177–1196 (2003)
30. Kikuchi, N., Nishiwaki, S., Fonseca, J.S.O., Silva, E.C.N.: Design optimization method for compliant mechanisms and material microstructure. Comput. Meth. Appl. Mech. Eng. **151**(3–4), 401–417 (1998)
31. Sigmund, O.: Materials with prescribed constitutive parameters: an inverse homogenization problem. Int. J. Solids Struct. **31**(17), 2313–2329 (1994)
32. Silva, E.C.N., Fonseca, J.S.O., Kikuchi, N.: Optimal design of periodic microstructures. Comput. Mech. **19**(5), 397–410 (1997)
33. Giusti, S.M., Novotny, A.A., de Souza Neto, E.A., Feijóo, R.A.: Sensitivity of the macroscopic elasticity tensor to topological microstructural changes. J. Mech. Phys. Solids **57**(3), 555–570 (2009)
34. Sokołowski, J., Żochowski, A.: On the topological derivative in shape optimization. SIAM J. Control Optim. **37**(4), 1251–1272 (1999)
35. Germain, P., Nguyen, Q.S., Suquet, P.: Continuum thermodynamics. J. Appl. Mech. Trans. ASME **50**(4), 1010–1020 (1983)
36. Amstutz, S., Giusti, S.M., Novotny, A.A., de Souza Neto, E.A.: Topological derivative for multi-scale linear elasticity models applied to the synthesis of microstructures. Int. J. Numer. Meth. Eng. 84, 733–756 (2010)
37. Novotny, A.A., Feijóo, R.A., Taroco, E., Padra, C.: Topological sensitivity analysis for three-dimensional linear elasticity problem. Comput. Meth. Appl. Mech. Eng. **196**(41–44), 4354–4364 (2007)

Topological Sensitivity Analysis for Two-Dimensional Heat Transfer Problems Using the Boundary Element Method

C. T. M. Anflor and R. J. Marczak

Abstract The objective of the current chapter is to present the application of a hard kill material removal algorithm for topology optimization of heat transfer problems. The boundary element method is used to solve the governing equations. A topological-shape sensitivity approach is used to select the points showing the lowest sensitivities, where material is removed by opening a cavity. As the iterative process evolves, the original domain has holes progressively introduced, until a given stop criteria is achieved. In a topological optimization process, final shapes with irregular boundaries are usual. Instead of applying boundary smoothing techniques at a postprocessing level, this work adopts a procedure in which smooth boundaries are ensured as a direct outcome of the original optimization code. The strategy employs Bézier curves for boundary parameterization. An algorithm is also developed to detect, during the optimization process, which curve of the intermediary topology must be smoothed. For the purpose of dealing with non-isotropic materials a linear coordinate transformation is implemented.

1 Introduction

The thermal conducting solid issue has received relatively little attention in spite of its significance for industrial applications such as polymer curing in die-molding processes, printed circuit board designs, and heat diffusers, to cite a few. Among the

C. T. M. Anflor (✉)
Campus Universitário Gama, Universidade de Brasília, Área Especial de Indústria Projeção A-UnB-72, Brasília, DF 444–240, Brazil
e-mail: anflor@unb.br

R. J. Marczak
Departamento de Engenharia Mecânica, Universidade Federal do Rio Grande do Sul, Rua Sarmento Leite 425, Porto Alegre, RS 90050–170, Brazil
e-mail: rato@mecanica.ufgrs.br

P. A. Muñoz-Rojas (ed.), *Optimization of Structures and Components*, Advanced Structured Materials 43, DOI: 10.1007/978-3-319-00717-5_2,

numerical techniques developed for topology optimization, solid isotropic material with penalization (SIMP) and evolutionary structural optimization (ESO) rank as the dominant methods, and both (along with their variants) have been successfully used in many optimization fields (a recent comprehensive review can be found in [1]. Classical topological optimization, based on the homogenization or the density approach [2, 3], often used for elasticity problems presents some drawbacks. One of these drawbacks refers to the final and intermediary topology from an optimization process, which results an appearance of sawtooth shape boundaries. This final shape irregularity frequently requires post-processing. Final shapes obtained by sensitivity analysis using topological derivative (D_T) [4] are frequently irregular too, because of the usual methods employed to remove material. This issue independs of the optimization method adopted.

Irregularities in the boundary shape, such as those resulting from the contour of rectangular cells, are not adequate for a realistic response analysis, since a non-physical field concentration shows up around sharp corners. In order to overcome this undesired effect, and following [5], a new methodology for topology optimization is implemented in the original code. The main idea consists in smoothing the boundary during the process of optimization, that is, shape and topology optimization are simultaneously performed in the design process. A smoothing technique based on the use of Bézier curves is implemented. Afterwards the final shape is compared with those obtained through the classical optimization solution. Figure 1 illustrates a BEM mesh smoothed by Bézier interpolation.

A number of linear heat transfer examples are solved with the formulation proposed. The irregular boundaries from the final and intermediaries shapes are eliminated. Materials with non-isotropic behavior are also considered by applying the linear coordinate transformation method. The final topologies are compared to those in the literature when available.

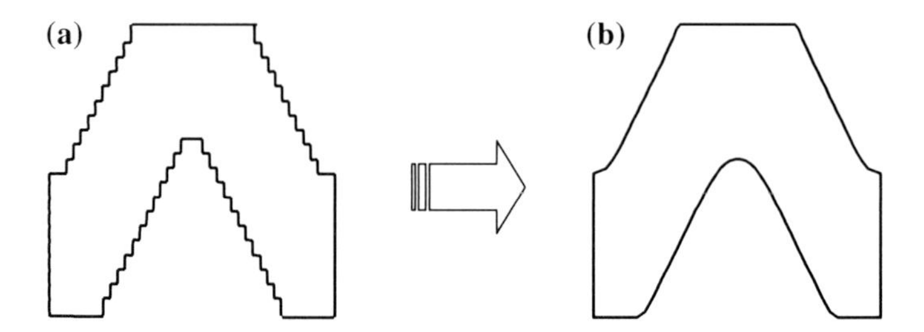

Fig. 1 Example of boundary smoothing. **a** Original result. **b** Beziér-smoothed result

2 Numerical Methodology

In this topic a schematic overview about all procedures implemented by the authors, and their sequences [6, 7] in the optimization code will be presented. The simplest BEM code is developed in its direct version using the fundamental solution for isotropic materials. The linear coordinate transformation method presented by Poon et al. [8] and Poon [9] is introduced in this routine. This technique makes it possible to solve anisotropic heat transfer problems, avoiding changes in the BEM code and manipulations of the D_T formulas. The governing differential equation for the heat conduction problem in a two-dimensional Cartesian coordinate system is given in its full form by

$$k_{11}\frac{\partial^2 T}{\partial x^2} + 2k_{12}\frac{\partial^2 T}{\partial x \partial y} + k_{22}\frac{\partial^2 T}{\partial y^2} = 0 \tag{1}$$

where, k_{11}, k_{12}, and k_{22} are the thermal conductivity coefficients, while T represents the temperature field. The corresponding heat fluxes are expressed as

$$\begin{aligned} q_x &= -k_{11}\frac{\partial T}{\partial x} - k_{12}\frac{\partial T}{\partial y} \\ q_y &= -k_{12}\frac{\partial T}{\partial x} - k_{22}\frac{\partial T}{\partial y} \end{aligned} \tag{2}$$

The initial geometry (x) established in an anisotropic medium is mapped into a corresponding geometry in an equivalent isotropic problem ($\hat{x}$). To do so, a special linear coordinate transformation is used, which transforms the partial differential equation into the Laplace equation, as

$$\begin{aligned} \hat{x} &= x + \alpha \cdot y \\ \hat{y} &= \beta \cdot y \end{aligned} \tag{3}$$

where $\alpha = \frac{-k_{12}}{k_{22}}$, $\beta = \frac{k}{k_{22}}$, $k = \sqrt{k_{11}k_{22} - k_{12}^2}$.

Neumann boundary conditions must also be transformed according to the Eq. (4).

$$\begin{aligned} q_y &= -k\frac{\partial T}{\partial y} = q_{\hat{y}} \\ q_x &= \beta q_{\hat{x}} - \alpha q_{\hat{y}} \end{aligned} \tag{4}$$

To demonstrate the steps of the numerical implementation, a numerical methodology scheme will be presented in details. The optimization process is carried out in 7 steps (see Fig. 2):

Step 1 Transform an orthotropic domain into an equivalent isotropic domain through the linear coordinate transformation expressed in Eq. (3). The heat flux is transformed by inverting Eq. (4).

Step 2 Solve the problems by means of the BEM code developed for isotropic materials. Evaluate D_T.

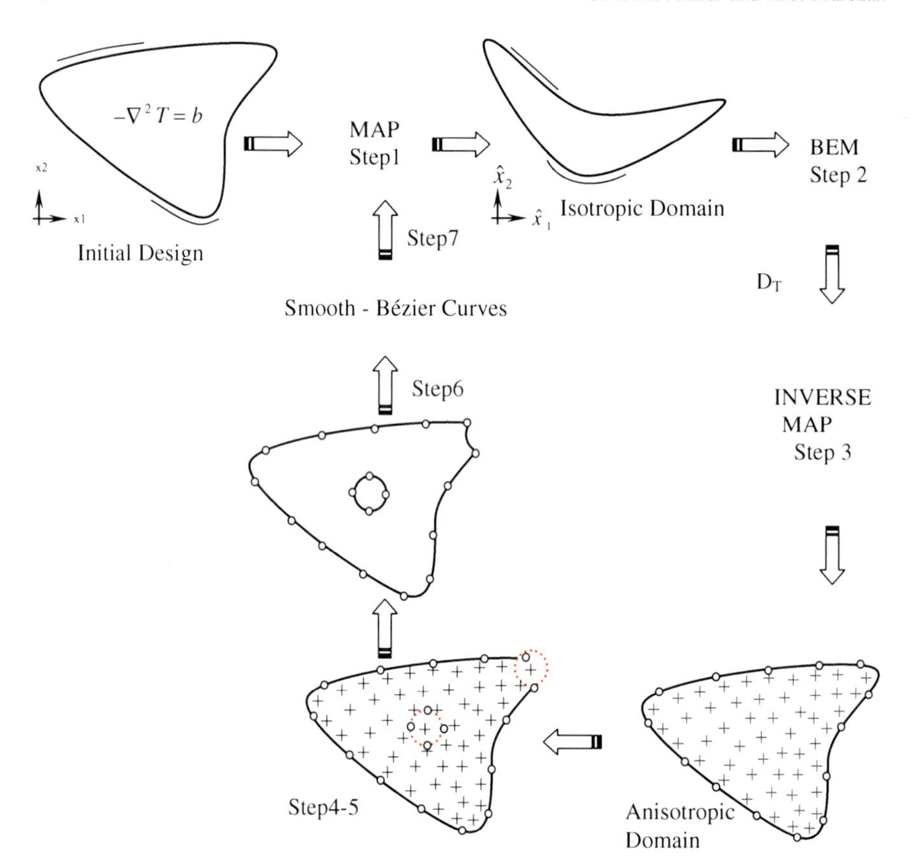

Fig. 2 Numerical methodology scheme

Step 3 Apply the inverse of the domain mapping using the equations available for the geometry and for the heat flux.
Step 4 The variables are evaluated on a suitable grid of interior points. The points with the lowest values of D_T are selected.
Step 5 Holes are created by "punching out" disks of material centered on the points previously selected.
Step 6 Smooth process using Bézier curves is introduced.
Step 7 Check stopping criteria, rebuild the mesh, return to step 1.

When the process is halted, a smooth final design topology of a non-isotropic material is expected. Note that in order to avoid geometrical boundary irregularities due to the algorithm used to remove material, a smoothing process is implemented at step 6. The smoothing technique uses Bézier curves to treat the polylines of intermediary geometries [10].

2.1 The Boundary Element Method

The formulation and application of the BEM for two-dimensional potential problems is very well established. In what follows only a brief description of the method is given. Further details can be found in [11] and [12]. Equation (5) presents the boundary integral equation which relates the potential u and flux q on the boundary Γ, in absence of body sources,

$$\frac{1}{2}u^i(x) + \int_{\Gamma} u(x)q^*(x, x')d\Gamma = \int_{\Gamma} q(x)u^*(x, x')d\Gamma \tag{5}$$

The functions u^* and q^* are the so-called potential and flux fundamental solutions due to a unit source applied at x'.

$$\begin{aligned} u^* &= \frac{1}{2\pi}\int_{\Gamma} \ln\left(\frac{1}{r}\right) d\Gamma \\ r &= \|x - x'\| \end{aligned} \tag{6}$$

The next step consists in discretizing the problem boundary Γ using N linear boundary elements, see Fig. 3.

The values of u and q at any point on an element can be written in terms of the nodal values and the two interpolation functions ϕ_1 and ϕ_2:

$$\begin{aligned} u(\zeta) &= \phi_1 u_1 + \phi_2 u_2 = [\phi_1 \phi_2] \begin{Bmatrix} u_1 \\ u_2 \end{Bmatrix} \\ q(\zeta) &= \phi_1 q_1 + \phi_2 q_2 = [\phi_1 \phi_2] \begin{Bmatrix} q_1 \\ q_2 \end{Bmatrix} \end{aligned} \tag{7}$$

where ξ is a local intrinsic coordinate defined in the range [−1,+1] and ϕ_1 and ϕ_2 are the standard discontinuous linear shape functions [11]. Considering the discretized version of Eq. (5), the integral in the left hand side over the element j can be written as,

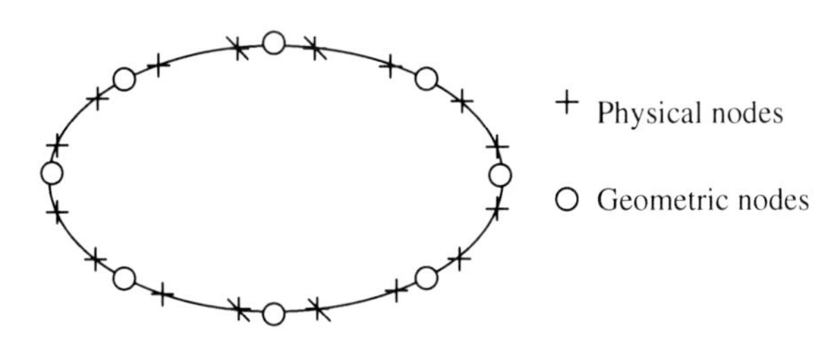

Fig. 3 Boundary element discretization

$$\int_{\Gamma j} uq^* d\Gamma = \int_{\Gamma j} [\phi_1 \phi_2] q^* d\Gamma \begin{Bmatrix} q_1 \\ q_2 \end{Bmatrix} = [h_1^{ij} h_2^{ij}] \begin{Bmatrix} q_1 \\ q_2 \end{Bmatrix} \tag{8}$$

where for each element j we have the two terms,

$$\begin{aligned} h_1^{ij} &= \textstyle\int_{\Gamma j} \phi_1 q^* d\Gamma \\ h_2^{ij} &= \textstyle\int_{\Gamma j} \phi_2 q^* d\Gamma \end{aligned} \tag{9}$$

Similarly, the integral on the right hand side results in

$$\int_{\Gamma_j} qu^* d\Gamma = \int_{\Gamma_j} [\phi_1 \phi_2] u^* d\Gamma \begin{Bmatrix} q_1 \\ q_2 \end{Bmatrix} = [g_1^{ij} g_2^{ij}] \begin{Bmatrix} q_1 \\ q_2 \end{Bmatrix} \tag{10}$$

where,

$$g_1^{ij} = \int_{\Gamma_j} \phi_1 u^* d\Gamma \text{ and } g_2^{ij} = \int_{\Gamma_j} \phi_2 u^* d\Gamma \tag{11}$$

After the substitution of Eqs. 11 and 9 for all the j elements in the discretized counterpart of Eq. 5, results:

$$c^i u^i + \sum_{j=1}^{N} H^{ij} u^j = \sum_{j=1}^{2N} G^{ij} q^j \tag{12}$$

After the imposition of all boundary conditions, the system in Eq. (12) can be reordered in such a way that all the unknowns are taken to the left hand side, resulting in the following system of equations:

$$[A]\{X\} = \{F\} \tag{13}$$

2.2 Topological Derivative

Topological derivative for Laplace and Poisson equations are applied in this work. A simple example of applicability consists in a case where a small hole of radius (ε) is open inside the domain. The concept of D_T consists in determining the sensitivity of a given cost function (ψ) when this small hole is increased or decreased (Fig. 4).

The local value of D_T at a point ($\hat{x}$) inside the domain for this case is evaluated by:

$$\overset{*}{D}_T(\hat{x}) = \lim_{\varepsilon \to 0} \frac{\psi(\Omega_\varepsilon) - \psi(\Omega)}{f(\varepsilon)} \tag{14}$$

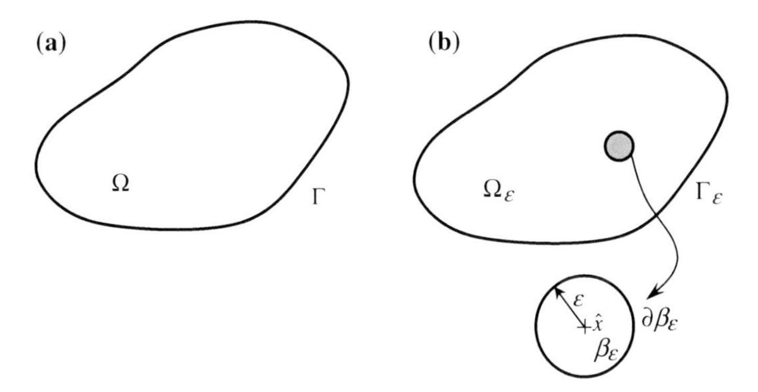

Fig. 4 Topological derivative original concept: **a** original domain Ω and **b** perturbed domain Ω_ε

where $\psi(\Omega)$ and $\psi(\varepsilon)$ are the cost function evaluated for the original and the perturbed domain, respectively, and is a problem dependent regularizing function. By Eq. (14) it is not possible to establish an isomorphism between domains with different topologies. This equation was modified introducing a mathematical idea that the creation of a hole can be accomplished by perturbing an existing one whose radius tends to zero, Fig. 5.

This allows the restatement of the problem in such a way that it is possible to establish a mapping and obtain an expression for the topological derivative, such that

$$\overset{*}{D}_T(\hat{x}) = \lim_{\varepsilon \to 0} \frac{\psi(\Omega_{\varepsilon+\delta\varepsilon}) - \psi(\Omega_\varepsilon)}{f(\Omega_{\varepsilon+\delta\varepsilon}) - f(\Omega\varepsilon)} \tag{15}$$

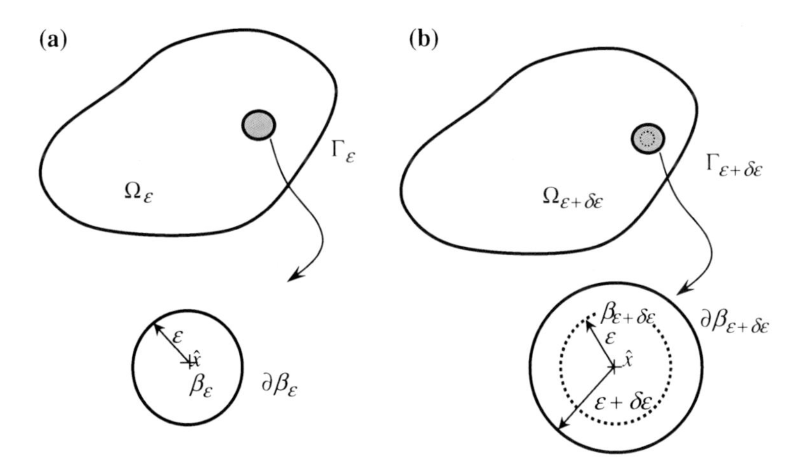

Fig. 5 Topological derivative with a new point of view. **a** Original domain and **b** perturbed domain

where δ_ε is a small perturbation on the holes's radius. Equation (15) gives a shape sensitivity when the hole becomes larger or smaller. It is also important to notice the both Eqs. 14, 15 are equivalent, as presented in [4]. The D_T for the steady state heat transfer will be briefly reviewed herein. In the case of linear heat transfer, the direct problem is stated as:

$$\text{Solve } \{u_\varepsilon | -k\Delta u_\varepsilon = b\} \text{ on } \Omega_\varepsilon \tag{16}$$

subjected to

$$\begin{cases} u_\varepsilon = \overline{u} & \text{on } \Gamma_D \\ k\frac{\partial u}{\partial n} = \overline{q} & \text{on } \Gamma_N \\ k\frac{\partial u_\varepsilon}{\partial n} = h_c\left(u_\varepsilon - u_\infty\right) & \text{on } \Gamma_R \end{cases} \tag{17}$$

where

$$h(\alpha,\beta,\gamma) = \underbrace{\alpha\left(u_\varepsilon - \overline{u}^\varepsilon\right)}_{Dirichlet b.c.} + \underbrace{\beta\left(k\frac{\partial u_\varepsilon}{\partial n} + \overline{q}^\varepsilon\right)}_{Neumann b.c.} + \underbrace{\gamma\left(k\frac{\partial u_\varepsilon}{\partial n} + h_c^\varepsilon\left(u_\varepsilon - u_\infty^\varepsilon\right)\right)}_{Robin b.c.} = 0 \tag{18}$$

is a function which takes into account the type of boundary condition on the holes to be created (u_ε, $\frac{\partial u_\varepsilon}{\partial n} = q_\varepsilon$ are the temperature and flux on the hole boundary, while and u_∞^ε and h_c^ε are the hole's internal convection parameters, respectively).

A general form for the cost function can be written as

$$\psi\left(\Omega_\tau\right) = \int\limits_{\Omega_\tau} \varphi_{\Omega_\tau}(u_\tau) d\Omega_\tau + \int\limits_{\Gamma_\tau} \varphi_{\Gamma_\tau}(u_\tau)\, d\Gamma_\tau \tag{19}$$

where τ is a parameter associated to the shape change velocity, i.e., $x_\tau(x) = x + \tau v(x)$. The sensitivity of the cost function with respect to τ can be derived from the Gâteaux derivative.

$$\frac{d}{d\tau}\Psi\left(\Omega_\tau\right)_{\tau=0} = \lim_{\tau\to 0}\frac{\Psi\left(\Omega_\tau\right) - \Psi\left(\Omega_{\tau|\tau=0}\right)}{\tau} h(\alpha,\beta,\gamma) = 0 \text{ on } \partial\beta_\varepsilon \tag{20}$$

Therefore the problem should be re-stated as

Evaluate: $\frac{d}{d\tau}\Psi\left(\Omega_\tau\right) = 0$

Subject to

$$a_\tau\left(u_\tau, n_\tau\right) = l_\tau\left(n_\tau\right) \qquad \forall n_\tau \in \beta_{\tau 1} \qquad \forall \tau \geq 0 \tag{21}$$

where a_τ is a continuous, coercive bilinear form, l_τ is a continuous linear functional and β_τ is the space of the admissible perturbation functions for the perturbed domain Ω_τ.Using the total potential energy as a cost function ($\Phi_\tau(u_\tau) := \frac{1}{2}a_\tau(u_\tau, u_\tau) - l(u_\tau)$), the bilinear form a_τ and the functional l_τ are written as:

$$a_{\varepsilon}\left(u_{\varepsilon}, n_{\varepsilon}\right) := \int_{\Omega_{\varepsilon}} k \nabla u_{\varepsilon} \cdot \nabla \eta_{\varepsilon} d\Omega + \int_{\Gamma_c} h_c u_{\varepsilon} \eta_{\varepsilon} d\Gamma + \int_{\partial \Lambda_{\varepsilon}} h_c^{\varepsilon} u_{\varepsilon} \eta_{\varepsilon} d\partial \Lambda$$

$$l_{\varepsilon}\left(\eta_{\varepsilon}\right) := \int_{\Omega_{\varepsilon}} b \eta_{\varepsilon} d\Omega - \int_{\Gamma} \bar{q} \eta_{\varepsilon} d\Gamma - \int_{\Gamma_c} h_c u_{\infty} \eta_{\varepsilon} d\Gamma - \int_{\partial \Lambda_{\varepsilon} \bar{q}_{\varepsilon}} \eta_{\varepsilon} d\partial \Lambda + \gamma \int_{\partial \Lambda_{\varepsilon}} h_c^{\varepsilon} u_{\infty} \eta_{\varepsilon} d\partial \Lambda \tag{22}$$

Considering Eq. (21) one can derive the D_T expression particularized for the three classical boundary conditions prescribed on the holes.

2.2.1 Neumann Boundary Condition

For this case, Eq. (9) is particularized as ($\alpha = 0, \beta = 1, \gamma = 0$) and the D_T is obtained by taking the limit:

$$D_T(\hat{x}) = -\lim_{\varepsilon \to 0} \frac{1}{2f'(\varepsilon)} \int_{\partial\Omega\varepsilon} \left[k\left(\frac{\partial u_{\varepsilon}}{\partial t}\right) - k\left(\frac{\partial u_{\varepsilon}}{\partial n}\right) - 2bu_{\varepsilon} - \frac{2}{\varepsilon}\bar{q}_{\varepsilon} u_{\varepsilon} \right] d\Omega\varepsilon \tag{23}$$

Both cases of Neumann boundary conditions must be considered:

$$\bar{q}_{\varepsilon} = \left.\frac{\partial u_{\varepsilon}}{\partial n}\right|_{\partial\Omega_{\varepsilon}} = 0 \text{ with } f'(\varepsilon) = -\pi\varepsilon^2 \tag{24}$$

$$\bar{q}_{\varepsilon} = \left.\frac{\partial u_{\varepsilon}}{\partial n}\right|_{\partial\Omega_{\varepsilon}} \neq 0 \text{ with } f'(\varepsilon) = -2\pi\varepsilon \tag{25}$$

For homogeneous and non-homogeneous cases respectively.

2.2.2 Dirichlet Boundary Condition

For this case, Eq. (9) is particularized as ($\alpha = 1, \beta = 0, \gamma = 0$) and the D_T is obtained by taking the limit:

$$D_T(\hat{x}) = -\lim_{\varepsilon \to 0} \frac{1}{2f'(\varepsilon)} \int_{\partial\Omega\varepsilon} \left[k\left(\frac{\partial u_{\varepsilon}}{\partial t}\right)^2 - k\left(\frac{\partial u_{\varepsilon}}{\partial n}\right)^2 - 2bu_{\varepsilon} \right] d\Omega\varepsilon \tag{26}$$

being the conditions

$$u_{\varepsilon} = \overline{u}_{\varepsilon} \qquad \left.\frac{\partial u_{\varepsilon}}{\partial t}\right|_{\partial\Omega_{\varepsilon}} \neq 0 \tag{27}$$

used along with $f'(\varepsilon) = \frac{-2\pi}{\ln \varepsilon}$.

Table 1 Topological derivative for the three b.c. prescribed on the holes

Boundary condition on the hole	Topological derivative	Evaluated at
Neumann homogeneous boundary condition ($\alpha = 0, \beta = 1, \gamma = 0$)	$D_T(\hat{x}) = k\nabla u \nabla u - bu$	$\hat{x} \in \Omega \cup \Gamma$
Neumann non-homogeneous boundary condition ($\alpha = 0, \beta = 1, \gamma = 0$)	$D_T(\hat{x}) = -q_\varepsilon u$	$\hat{x} \in \Omega \cup \Gamma$
Robin boundary condition ($\alpha = 0, \beta = 0, \gamma = 1$)	$D_T(\hat{x}) = h_c^\varepsilon (u_\varepsilon - u_{\varepsilon\infty})$	$\hat{x} \in \Omega \cup \Gamma$
Dirichlet boundary condition ($\alpha = 1, \beta = 0, \gamma = 0$)	$D_T(\hat{x}) = -\frac{1}{2}k(u - \bar{u}_\varepsilon)$	$\hat{x} \in \Omega$
Dirichlet boundary condition ($\alpha = 1, \beta = 0, \gamma = 0$)	$D_T(\hat{x}) = k\nabla u \nabla u - b\bar{u}_\varepsilon$	$\hat{x} \in \Gamma$

2.2.3 Robin Boundary Condition

In this case one has ($\alpha = 0, \beta = 0, \gamma = 1$) and the D_T is obtained by taking the limit:

$$D_T(\hat{x}) = -\lim_{\varepsilon \to 0} \frac{1}{2f'(\varepsilon)} \int_{\partial\Omega\varepsilon} \left[k\left(\frac{\partial u_\varepsilon}{\partial t}\right)^2 - k\left(\frac{\partial u_\varepsilon}{\partial n}\right)^2 - 2bu_\varepsilon - \frac{2}{\varepsilon} h_c^\varepsilon (u_\varepsilon - 2u_{\varepsilon\infty}) \right] d\Omega\varepsilon \tag{28}$$

Now the regularizing function is $f'(\varepsilon) = -2\pi\varepsilon$.

Table 1 summarizes the final expressions for D_T after application of the respective regularizing functions for each boundary condition, in accordance to Eqs. (23), (26) and (28).

It is also important to take attention that D_T is evaluated by different expressions for internal and boundary points. Another remark relies on the fact that the expressions presented in Table 1 are deduced taking the total potential energy as cost function. Once the internal points have been calculated by BEM (at step 2 in Fig. 2) one can determine the domain's sensitivity by using the equations presented in Table 1. It is important to note that the D_T is calculated in the transformed domain $(\hat{x}, \hat{y})$ where its behavior is isotropic and the material is removed in the original domain (x, y) between steps 4-5, as depicted in Fig. 2.

2.3 *Bézier Curves*

Generally in an optimizations process the final topology results in a non-smooth geometry and requires the use of smoothing techniques during the optimization process. The most popular techniques to deal with these irregular geometries employ Bézier curves (adopted in this work), Douglas-Peucker and B-Splines algorithms. The Bézier curve approach was pioneered by Renault for modeling of surfaces in automobile design [10]. Bézier defines the curve *P(u)* in the terms of the location of $n + 1$ control points p_i.

$$P(u) = \sum_{i=0}^{n} p_i B_{i,n}(u) \tag{29}$$

where $B_{i,n}(u)$ are blending functions or polynomials of Bernstein

$$B_{i,n}(u) = C(n,i)u^i(1-u)^{n-i} \tag{30}$$

and $C(n,i) = \frac{n!}{i!(n-i)!}$ are the binomial coefficients.

Here, Eq. 29 is a vector and could be expressed by writing equations for the x and y parametric functions separately:

$$\begin{aligned} x(u) &= \sum_{i=0}^{n} x_i B_{i,n}(u) \\ y(u) &= \sum_{i=0}^{n} y_i B_{i,n}(u) \end{aligned} \tag{31}$$

x_i and y_i are the coordinates of the control points of the curve, always n + 1 points. The union of these points form the vertices of the control polygon of the Bézier curve. These points are responsible to control the shape of the curve, with the parameter u varying between 0 to 1. Further details can be found in [13]. As exposed until here the techniques for smoothing curves are well established in the literature and there is no difficulty associated to their application. However, in an optimization problem the effort relies on identifying which portions of the intermediary topologies must be smoothed. There are some parts of the topology that cannot be smoothed, such as the portion with prescribed boundary conditions or the portion which is a straight line. In order to overcoming this problem a routine is developed in the present work to identify, during an iterative optimization process, which curves must be or not smoothened. This routine was introduced inside the optimization algorithm as step 6, after the step of material removal (see Fig. 2). Figure 6 depicts the scheme of identification and smoothing of curves that takes place during the optimization process. This is a subroutine implemented between step 5 and 7 of the original code, as illustrated in Fig. 2.

3 Numerical Results

This section presents some examples that demonstrate the application of the proposed method. The results obtained for the first example are compared to those obtained by Park [14] for isotropic materials. The second example differs from the first one only in the boundary conditions, which prescribe convection in the cavities. The third and fourth examples consist in a square domain under high and low temperature boundary

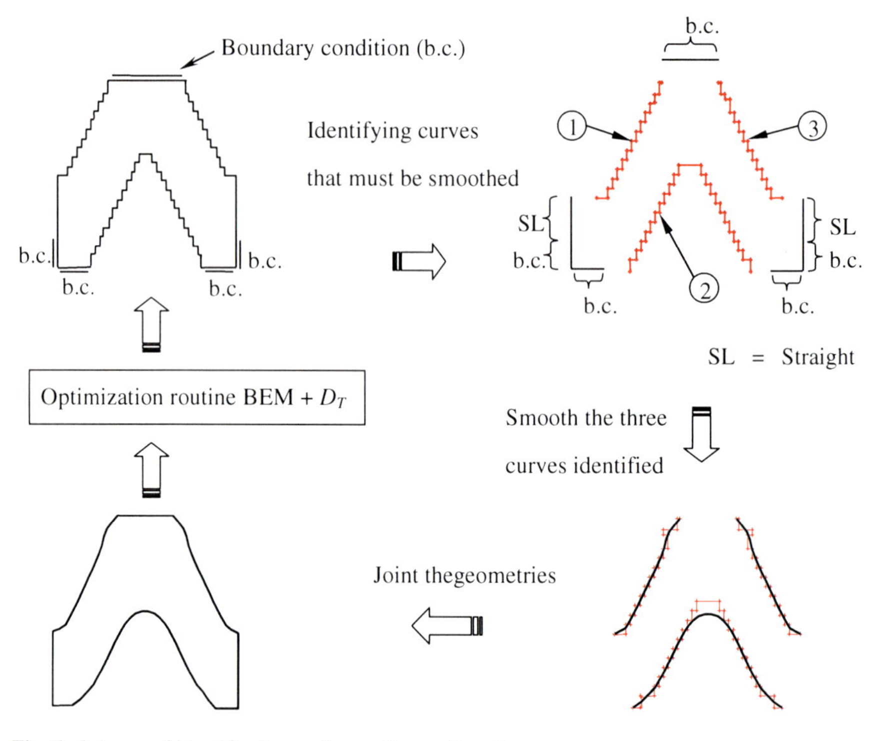

Fig. 6 Scheme of identification and smoothness–Step 6

conditions where the constitutive properties are varied to result in all possible material behaviors: isotropic, orthotropic and anisotropic. Finally, the last example considers an isotropic domain with heat sources. For this case the mapping technique and the smoothing process were not employed. The history of material removal is analyzed and illustrated for each case. The iterative process is halted when a given amount of material is removed from the original domain, regardless of the type of material medium. This criterion provides a basis to compare the topologies generated for isotropic, orthotropic and anisotropic media under the same initial geometry and boundary conditions. In all cases the total potential energy is used as the cost function. A regularly-spaced grid of internal points is generated automatically, taking into account the radius of the holes created during each iteration. The radius is obtained as a fraction of a reference dimension of the domain ($r = \omega l_{ref}$). In all cases $l_{ref} =$ min(height,width) is adopted. The objective in all cases is to minimize the material area. The current area of the domain (A_f) is checked at the end of each iteration until a reference value is achieved ($A_f = \varphi A_0$, where A_0 represents the initial area and φ a defined percentage of material to be removed). After that, the intermediary topology is smoothed by using Bézier functions. Linear discontinuous boundary elements integrated with 6 Gauss points are used in all cases.

3.1 Heat Conductor with Neumann Boundary Conditions on the Cavities

A rectangular 20×30 units domain subjected to prescribed temperature ($T_1 = 393\,\text{K}$) on its left edge and convection boundary conditions ($T_0 = 298\,\text{K}$ *and* $h_0 = 5.677\,\text{W/m}^2\,\text{K}$) on the remaining ones is to be optimized (Fig. 7). Here, the problem is revisited using only isotropic material properties. The isotropic material used is Aluminum ($k = 236\,\text{W/mK}$). For this case, Neumann boundary conditions are prescribed on the cavities open during the optimization process.

Figure 8 shows the evolution history obtained until the final area reached 30 % of the original value. The final and intermediary topology result in a smooth appearance shape. It is important to note that the appearance of saw-tooth shape on boundaries is avoided, discarding a post-processing in a manufacturing process.

The mean flux on the left edge side of the plate is chosen to take into account the behavior as the process evolves. The values of the mean flux obtained during the process of optimization including Bézier smoothing are recorded and compared to the result obtained with the original code with no smoothing. These results are depicted in Fig. 9 where is possible to see the evolutive iteration × mean flux. Also, in the same figure, the smoothed intermediary topology is depicted for some iterations.

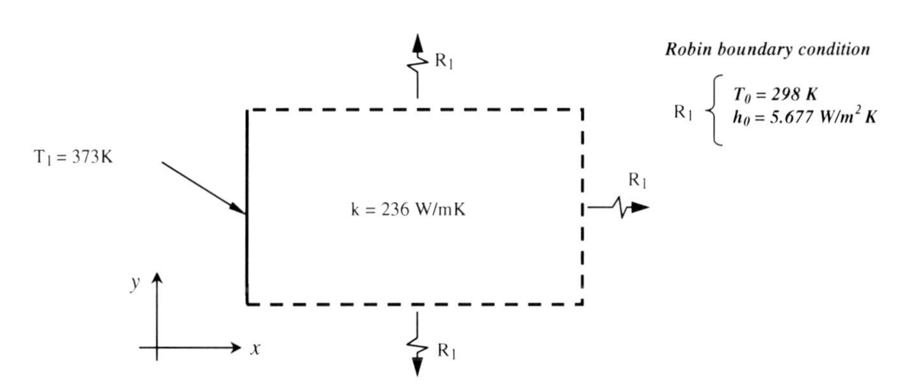

Fig. 7 Initial design domain

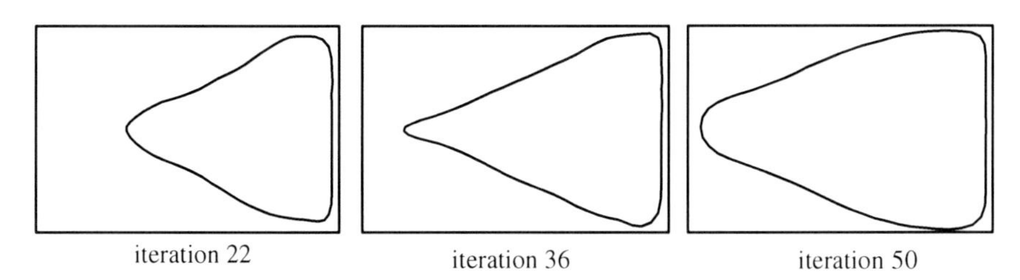

Fig. 8 Evolution history for isotropic media

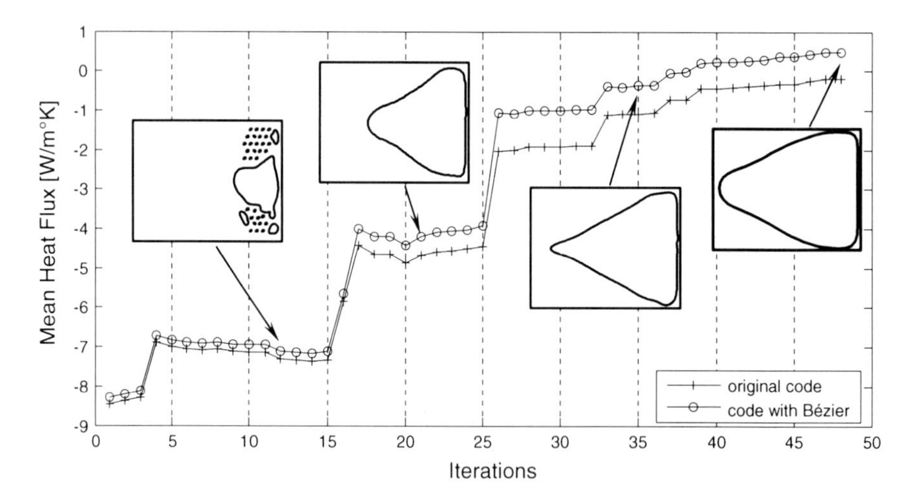

Fig. 9 Optimization history of mean heat flux versus iteration

In the orthotropic case, the thermal conductivities are imposed as $k_y/k_x = 2$. Figure 10 presents the evolution history obtained until the same volume ratio of the isotropic case was reached. Clearly, the resulting geometry of the internal cavity has a more pronounced curvature, so as to ease the heat flux in the y direction.

Park [14] solved this problem by using homogenization techniques and the FEM. Figure 11 compares the results obtained by Park [14] with the ones obtained with

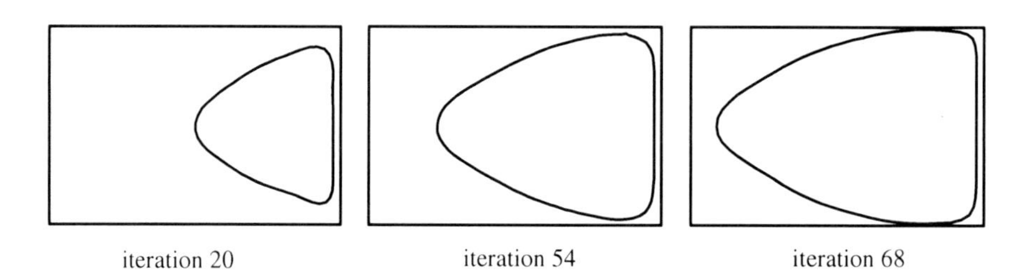

Fig. 10 Evolution history for orthotropic media

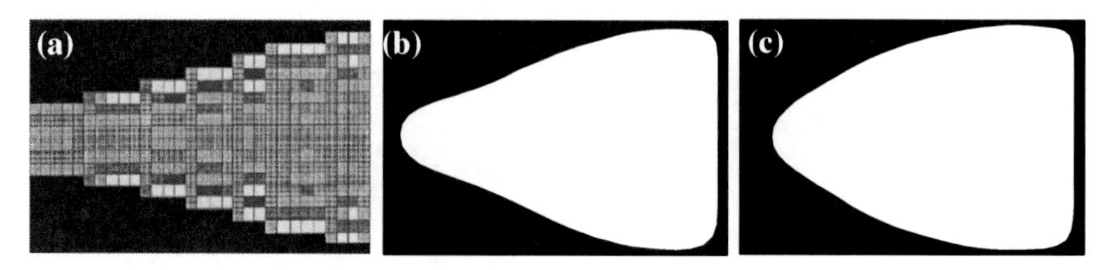

Fig. 11 Final topologies: **a** FEM+homogeneization [14] – $k_x/k_y = 1$; **b** Present work, BEM+ $D_T - k_x/k_y = 1$; **c** Present work, BEM+ $D_T - k_y/k_x = 2$

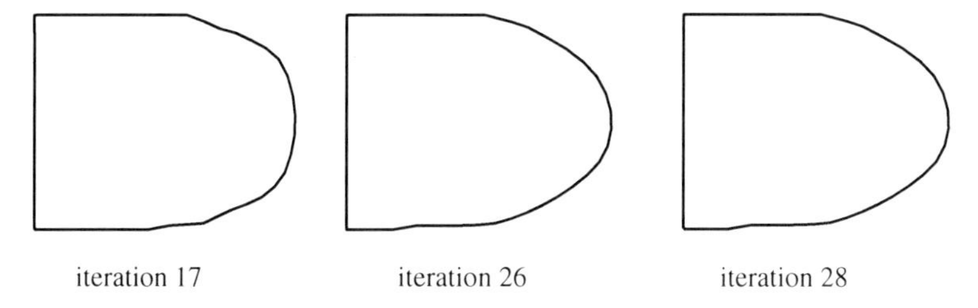

Fig. 12 Evolution history under Robin boundary conditions

the present method. In all cases, the final geometry is satisfactorily leading to a high-conductivity layout, and both isotropic results match.

3.2 Heat Conductor with Robin Boundary Conditions on the Cavities

This case is very similar to the previous one, except for the fact that the convection boundary condition is imposed on the cavities. This is an isotropic case optimized to a volume constraint of 65 % of the original design domain. The evolution of the optimization process is depicted in Fig. 12. It is important to point out that the final geometry converged to the shape of an optimal fin.

3.3 Inverted V Heat Conductor

This example consists in a square domain with high temperature (373 K) applied to its lower corners, while a low temperature (273 K) is applied at the mid top edge. The remaining boundaries are insulated. The cavities are created with $r = 0.04\, l_{ref}$ and the process is halted when $A_f = 0.6A_0$ is attained. For the purpose of illustrating and comparing the final topologies obtained, three variations of the present example are studied as:

$$\begin{aligned}
&\text{Case A}: \quad && k_{xx} = 1; \quad && k_{yy} = 1 \\
&\text{Case B}: \quad && k_{xx} = 2; \quad && k_{yy} = 1 \\
&\text{Case C}: \quad && k_{xx} = 3; \quad && k_{yy} = 1
\end{aligned}$$

Figure 13 shows the results for the isotropic case (case A) which is used to compare the final design with those of the orthotropic cases (cases B and C). Figures 14 and 15

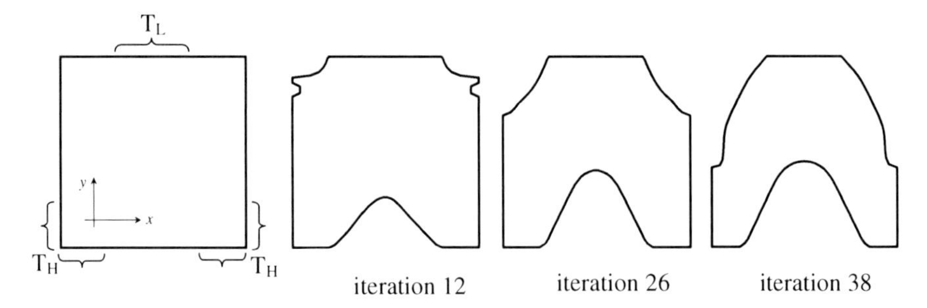

Fig. 13 Evolution history for isotropic material—Case A

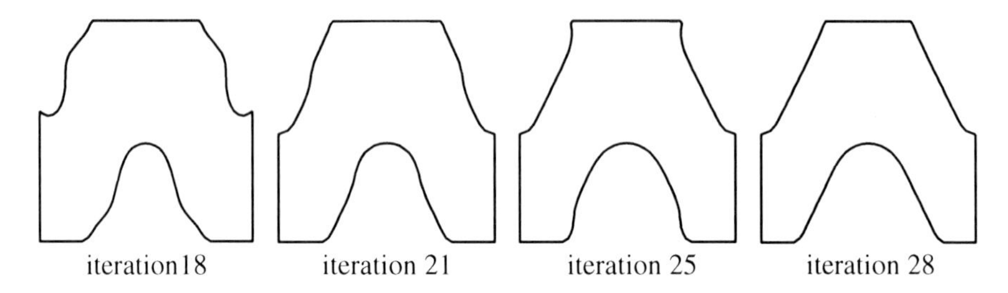

Fig. 14 Evolution history for orthotropic material—Case B

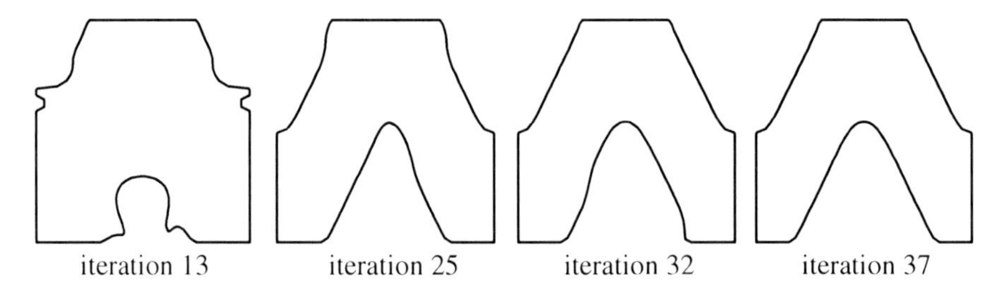

Fig. 15 Evolution history for orthotropic material—Case C

present the optimization evolution for the orthotropic cases and their final topologies when the stop criteria is achieved. From this it is possible to compare the three cases. There are visible differences in the evolution of material removal for each case. Therefore, the final designs are slightly different.

Figure 16 presents the evolution of material removal for all cases. It is found that highly orthotropic cases result in higher values of the topological sensitivity, in comparison to the isotropic solution. Consequently, a larger material removal rate is expected for orthotropic problems, in general, but this is an assertion that highly depends on the nature of the problem.

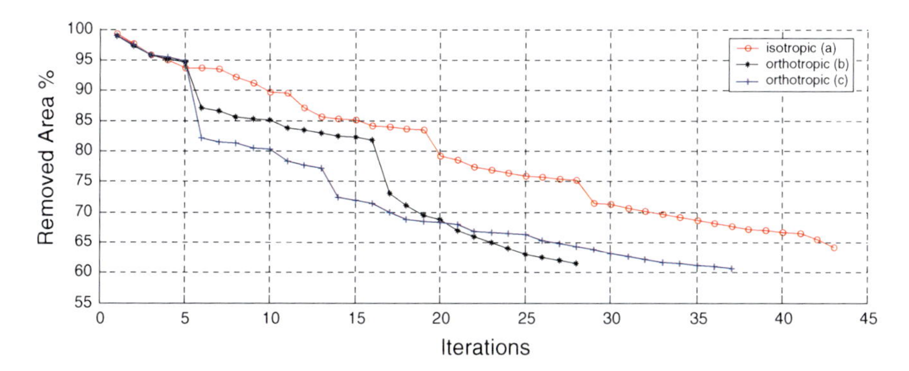

Fig. 16 Material removal history for the inverted V heat conductor

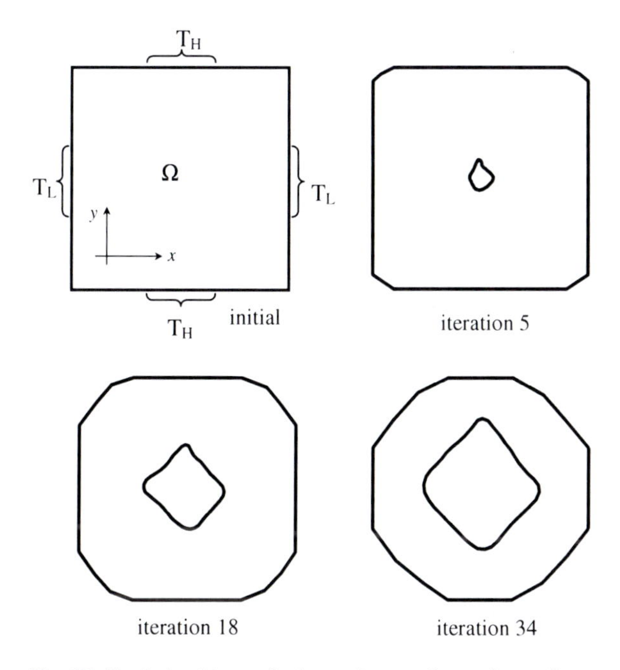

Fig. 17 Evolution history for isotropic case ($k_{11} = k_{22} = 1$)

3.4 Cross Heat Conductor

This example refers to a square domain subjected to low and high temperature boundary conditions on the middle of opposite sides. The problem is depicted in Fig. 17, where T_H is the high temperature (373 K) and T_L is the low temperature (273 K). The remaining boundaries are insulated. All possible cases are studied: isotropic, orthotropic and anisotropic materials and they are to be optimized until $A_f \approx 0.4 A_0$ is achieved.

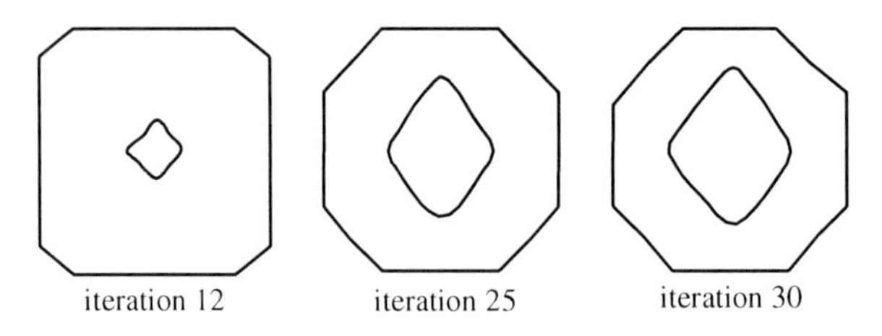

Fig. 18 Evolution history for orthotropic case ($k_{xx}/k_{yy} = 5$)

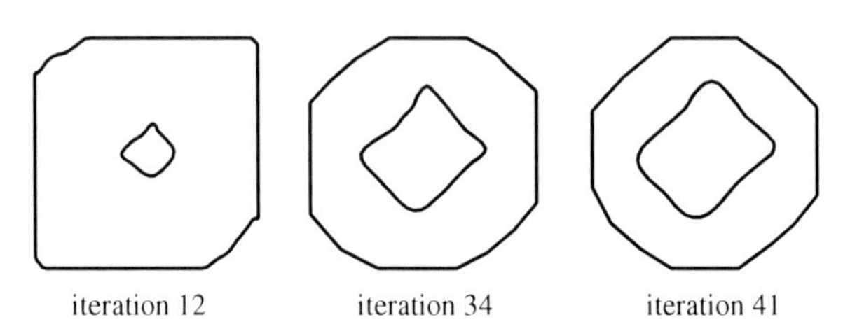

Fig. 19 Evolution history for anisotropic case ($k_{xx} = 1$, $k_{yy} = 1$; $k_{xy} = 0.5$)

Initially, an isotropic case is analyzed with $k_{11} = k_{22} = 1$. Symmetry is not used to provide a direct comparison to the subsequent anisotropic cases (which cannot use symmetry). Figure 14 also shows the evolution of material removal for $r = 0.02l_{ref}$. It is important to observe that the algorithm delivers fairly symmetric solutions throughout the process. The condition $A_f \approx 0.4 \cdot A_0$ is achieved after 34 iterations. The second case represents a highly orthotropic material, with the conductivities set to $k_{xx} = 5$ and $k_{yy} = 1$ (see Fig. 18). As expected, material is selectively removed so that the heat flux along the x direction is increased. The stop criterion $A_f \approx 0.4A_0$ is achieved after 30 iterations.

The third case considers an anisotropic material with $k_{xx} = 1$, $k_{yy} = 1$ and $k_{xy} = 0.5$. The evolution history is presented in Fig. 19, showing that the initial symmetry is lost after the first iterations, as expected. Contrary to the previous cases, the internal cavity results in a rhombic shape since the Cartesian axes are not parallel to the main axes of the constitutive matrix.

Figure 20 shows the percentage of material removed as a function of the number iterations for each case studied in the cross heat conductor. All cases are stopped when about 40 % of material is removed. These cases are analyzed without the aid of symmetry, for comparison purposes. Obviously, anisotropic cases cannot use symmetry in general, but in many practical situations it is possible (or even expected) to align the axes of the component with the principal directions of the constitutive matrix. In such cases, smoother designs can be obtained. In order to provide a further benchmark, the cross heat conductor example is re-analyzed for the isotropic and orthotropic cases using only one quadrant of the original geometry.

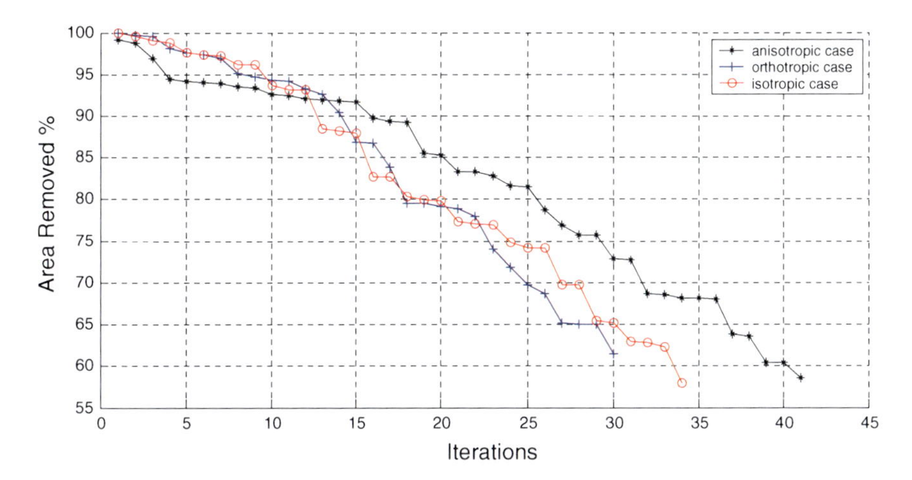

Fig. 20 Material removal history for the cross heat conductor

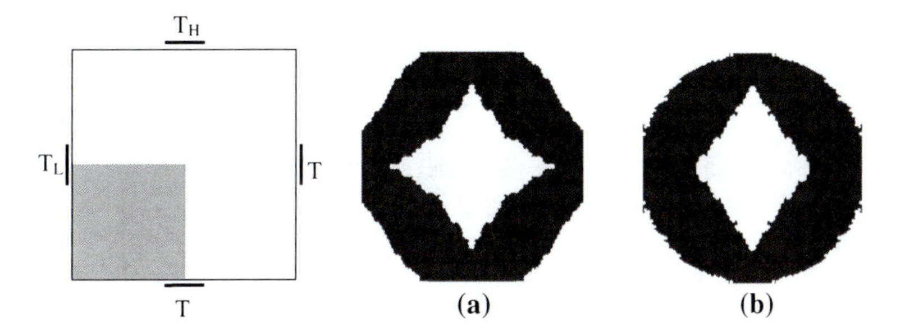

Fig. 21 Final topologies for: **a** isotropic and **b** orthotropic examples

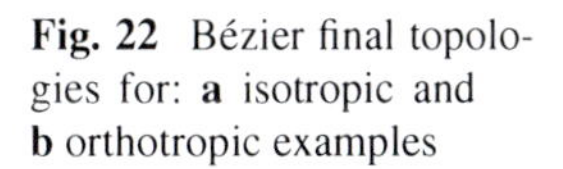

Fig. 22 Bézier final topologies for: **a** isotropic and **b** orthotropic examples

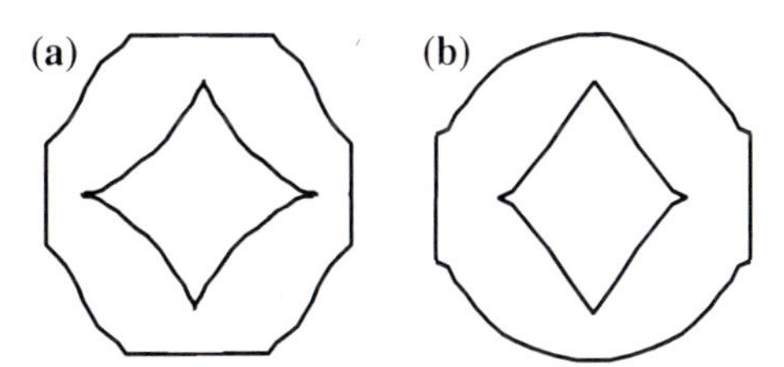

Figure 21 shows the final topologies obtained for both cases while Fig. 22 depicts the same topology after the smoothing process. The material is removed initially with $r = 0.04\ l_{ref}$ and then $r = 0.02\ l_{ref}$. for the remaining iterations. This simple expenditure helps generating smoother boundaries in the final design.

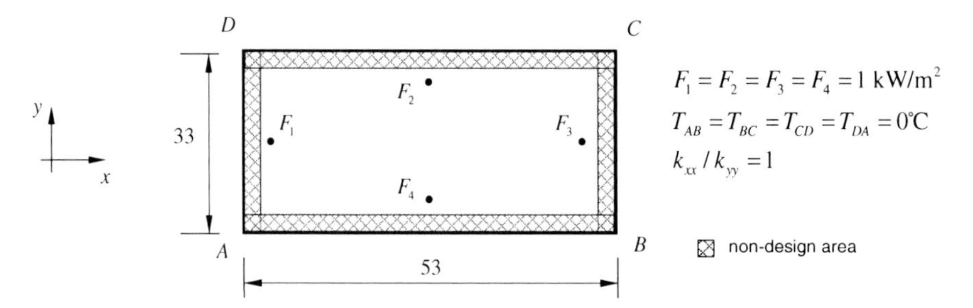

Fig. 23 PCB: initial design

3.5 Printed Circuit Board

This example was proposed by Li et al. [15] and refers to a printed circuit board (PCB) substrate. An important aspect of good PCB designs is the ability to dissipate the maximum amount of thermal energy with the minimum possible volume of material. Topology optimization is a valuable tool in such cases, allowing material savings unfeasible using only intuitive design ideas. Figure 23 shows the initial layout for this case, where four heat sources are used to simulate the heat generated by major electronic components mounted on the PCB. Part of the domain cannot be changed (hatched areas in Fig. 23). The BEM model used for this case employs an initial mesh of 32 elements linear elements, and the internal points are placed only within the design area. All the cavities have prescribed Neumann homogeneous boundary condition. The process is halted when a volume ratio of 70 % between the final and the original designs is achieved. Figure 24 shows the topology history. It is worth to note that all previous examples could be solved by pure shape optimization, since the original topology is not changed. In the PCB case, however, new cavities are created during the process, characterizing truly topological changes in the domain. It is also interesting to note the creation of small cavities near the corners of the PCB. The present approach shows good agreement with the results obtained by Li et al. [15], as shown in Fig. 25.

4 Additional Comments

The outcomes stressed in the previous section show clearly that, although useful results can be obtained with the proposed methodology, it shares the same problems of most hard kill methods. If conveniently controlled, the designs obtained may be acceptable. However, it may be impossible to specify a sufficiently general material removal rule regardless the application. Furthermore, the imposition of constraints to the problem is rather complex, since the analytical expressions for D_T have been derived (in the original references) without considering them.

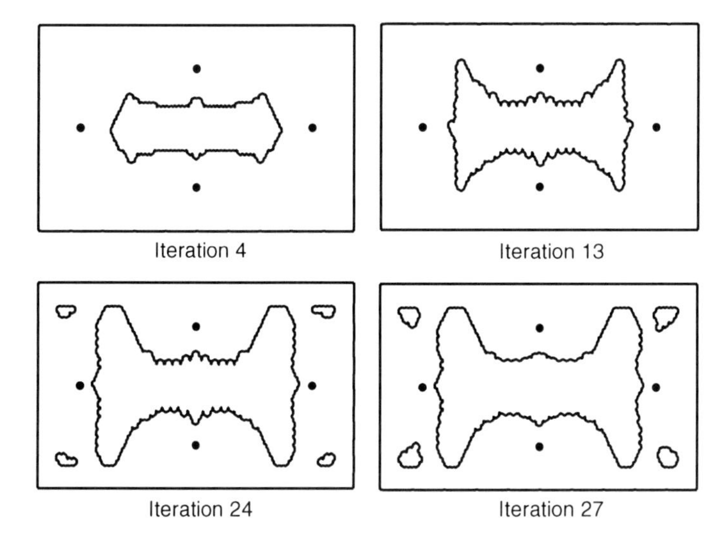

Fig. 24 Topology evolution for the PCB

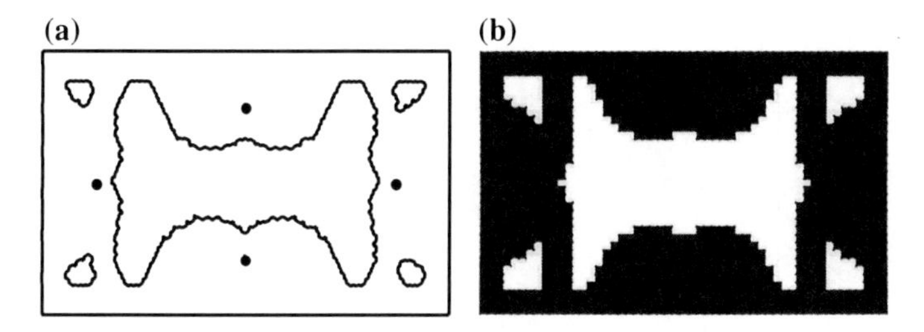

Fig. 25 Final topologies for the PCB: **a** Present results $-A_f = 0.70A_0$; **b** [15] $-A_f = 0.68A_0$

An important point in the present formulation is related to the relative magnitude of the D_T values as the iterative process evolves. After the first few iterations, the gradient of D_T along the domain becomes very low, since the optimal solution has constant sensitivities. Since some areas of the domain may be more affected by modeling and discretization errors than others, a fine tuning becomes mandatory to avoid material rejection at non-optimal locations. Another issue is related to the (non-mathematical) stopping criteria used. Specified volume fractions are relatively easy to implement. In the present approach, the current volume is compared with the target volume at the end of each iteration, and the process is halted if necessary (this is possible because the number of holes and the diameters are known during run time). Although useful for material saving, this criteria is devoid of further information like energy content or compliance values.

If, by one side, ESO methods show serious limitations, on the other side one can make interesting usages of D_T, particularly in conjunction with other optimization methods. For instance, if a surface of D_T values is constructed, its isolines can be used to specify a material rejection criteria. The domain obtained after that first iteration can be a good starting point for SIMP methods or a shape optimization analysis. Generally the algorithm used to remove material leaves something to be desired, delivering unpleasantly jagged boundaries. "Tiling" schemes of material removal [5] has already proved to generate solutions very similar to the ones obtained with SIMP methods, and can be alternatively used. In any case, the smoothness of the resulting boundaries is more related to the size of the holes used to remove material. Regardless the type of hole (circular or squared), the resolution of boundaries generated will be finer when clouds of smaller holes are punched out of the design domain. However, this means a large number of internal points where D_T must be evaluated, increasing the computational cost of the analysis. The ideal situation relies on using holes large enough to accelerate the material removal but yet sufficiently small to capture all relevant geometric features of the optimal design. The sensitivity analysis is not directly affected by the size of the elements, provided the mesh is no too coarse (i.e. providing good results on internal points). The important aspect here is to keep the relative magnitudes of D_T reasonably well estimated along the whole design domain.

Another point to be highlighted here is the possibility of using post-processing algorithms to smooth the boundaries obtained not only with the BEM methodology shown here, but also to eliminate the see-saw aspect of most SIMP approaches reported. In this chapter a Bézier curve approach is employed avoiding boundaries irregularities. It is striking to note that the use of Bézier curves discards the need of a very fine mesh of internal grid points (very small holes).

5 Conclusions

Procedures for topology optimization are object of present research interest and several techniques have been proposed during last years. Despite the evolution of the present techniques of optimization, many drawbacks specific to each method still must be fixed. In this regard, this work employs topological sensitivity analysis and boundary elements for delivering optimized topologies as an alternative to the traditional optimization methods. Complex shapes with irregularities on their boundaries frequently result after a process of topology optimization. In this work, a Bézier smoothing technique is employed in order to avoid a new task of post-processing over the resulting topology due to the boundary irregularities. This procedure allows attaining more realistic geometries when the optimization is halted. The use of this technique provides optimized geometries suitable for direct manufacturing of the final design without major designer interference. A linear coordinate transformation method is also implemented, allowing the use of a D_T expression originally formulated only for isotropic materials also for non-isotropic materials. It is important to point out that the D_T has the potential total energy as an implicit cost function. The linear heat transfer for Laplace and Poisson problems are solved showing the feasibility of the procedure proposed and agreement with other solutions.

References

1. Mackerle, J.: Topology and shape optimization of structures using FEM and BEM: a bibliography (1999–2001). Finite Elem. Anal. Des. **39**, 243–253 (2003)
2. Bendsøe, M.P., Kikuchi, N.: Generating optimal topologies in structural design using a homogenization method. Comput. Meth. Appl. Mech. Engrg. **71**, 197–224 (1998)
3. Hassani, B., Hinton, E.: A review of homogenization and topology optimization III: topology optimization using optimality criteria. Comput. Struct. **69**, 739–756 (1998)
4. Novotny, A., Feijóo, R., Taroco, E., Padra, C.: Topological-shape sensitivity analysis. Comput. Meth. Appl. Mech. Engrg. **192**, 803–829 (2003)
5. Marczak, R.J.: Topology optimization and boundary elements: a preliminary implementation for linear heat transfer. Eng. Anal. Bound. Elem. **31**, 793–802 (2007)
6. Anflor, C.T.M., Marczak, R.J.: A boundary element approach for topology design in diffusive problems containing heat sources. Int. J. Heat Mass Transf. **52**, 4604–4611 (2009)
7. Anflor, C.T.M., Marczak, R.J.: Topological optimization of anisotropic heat conducting devices using Bézier-smoothed boundary representation. Comput. Model. Eng. Sci. 78, 151–168 (2011) (Print).
8. Poon, K.C., Tsou, R.C.H., Chang, Y.P.: Solution of anisotropic problems of first class by coordinate-transformation. J. Heat Transf. **101**, 340–345 (1979)
9. Pooń, K.C.: Transformation of heat conduction problems in layered composites from anisotropic to orthotropic. Lett. Heat Mass Transf. **6**, 503–511 (1979)
10. Newman, W.M., Sproull, R.F.: Principles of Interactive Computer Graphics. McGraw-Hill, New York (1982)
11. Brebbia, C.A., Dominguez, J.: Boundary Elements: An Introductory Course. Computational Mechanics Publications, Southampton (1988)
12. Banerjee, P.K.: The Boundary Element Methods in Engineering. McGraw-Hill College, New York (1994)
13. Harrington, S.: Computer Graphics: A Programming Approach. McGraw-Hill, New York (1983)
14. Park, Y.K.: Extensions of optimal layout design using the homogenization method. Ph.D. Thesis, The University of Michigan, East Lansing Michigan (1995).
15. Li, Q., Steven, G., Querin, O., Xie, Y.: Shape and topology design for heat conduction by evolutionary structural optimization. Int. J. Heat Mass Transf. **42**, 3361–3371 (1999)

Design of Compliant Mechanisms with Stress Constraints Using Topology Optimization

Luís Renato Meneghelli and Eduardo Lenz Cardoso

Abstract Compliant mechanisms are mechanical devices that transform or transfer motion, force or energy through a single part. These mechanisms have important applications in micro electromechanical systems (MEMS) and other systems that require great accuracy in motion and micro scale. The compliant mechanisms design is performed by Topology Optimization Method, and the optimization problem is formulated to maximize strain-energy stored by mechanism, eliminating the appearance of hinges. The kinematic behavior of the mechanism is imposed through a set of constraints over some displacement degrees of freedom of interest. The elastic behavior of the compliant mechanisms is imposed using a global stress constraint and some important issues associated to stress parametrization are discussed in the realm of mechanism design. The characteristics and the feasibility of this proposal, as well as the influence of parameters related to the formulation, are presented with the aid of some examples.

1 Introduction

The topology optimization method aims to find a suitable distribution of a base material in a fixed sized domain, Ω, in order to attain the extreme of a given constrained functional [1]. This method has been used in a variety of problems, including the design of stiffer mechanical structures, heat transfer, wave propagation and material design. When dealing with the distribution of an elastic isotropic material, the Topology Optimization method consists in establish a relation, known as parametrization, between the property of a base material $\mathbf{T}^0$ and the property at a given point X inside

L. R. Meneghelli · E. L. Cardoso (✉)
State University of Santa Catarina, Joinville, SC, Brazil
e-mail: lrmeneghelli@gmail.com

E. L. Cardoso
e-mail: lenz@joinville.udesc.br

P. A. Muñoz-Rojas (ed.), *Optimization of Structures and Components*,
Advanced Structured Materials 43, DOI: 10.1007/978-3-319-00717-5_3,

Ω, $\mathbf{T}(X)$, in the form

$$\mathbf{T}(X) = \rho^p(X)\mathbf{T}^0 \tag{1}$$

where $\rho(X)$ is know as (pseudo) density and p is an exponent used to adjust Eq. 1 to a suitable behavior. If the domain is discretized using finite elements, an usual approach is to consider that each element is composed of a single material, such that Eq. 1 can be written as

$$\mathbf{T}_e = \rho_e^p(X)\mathbf{T}^0 \tag{2}$$

where the index e is related to the element. Using the material parametrization depicted in Eq. 2, the stiffness matrix of a compatible finite element can be written as

$$\mathbf{K}_e = \int_{\Omega_e} \mathbf{B}^T\mathbf{T}_e\mathbf{B}\ \ d\Omega_e = \rho_e^p \int_{\Omega_e} \mathbf{B}^T\mathbf{T}_e^0\mathbf{B}\ \ d\Omega_e = \rho_e^p\mathbf{K}_e^0 \tag{3}$$

where $\mathbf{K}_e^0$ is the stiffness matrix of element e considering the base material.

A very important application of the topology optimization method is the design of compliant mechanisms, where a single elastic part is used to transfer force, motion or energy from one point in space to another point, as depicted in Fig. 1.

There are many proposals to cope with such task, as for example, the maximization of the output displacement delivered by the mechanism [14], multi objective optimization [6, 9] or the maximization of the elastic energy stored inside the domain [3]. A common drawback associated to many of such formulations is the appearance of one-node connected finite elements, known as hinges. This hinge is not compatible with the basic requirement of a flexible mechanism, since it must be made by a single part. Also, as there is no rotational degree of freedom in classical elasticity, this movement is not associated to an internal elastic energy. There are many proposals in the literature to address this issue, as for example [12], where the appearance of hinges is associated to a geometrical constraint in the optimization problem and [15], where projection operators are used. Cardoso and Fonseca, [3], presented an alternative approach, where the main objective is changed. Instead of using another

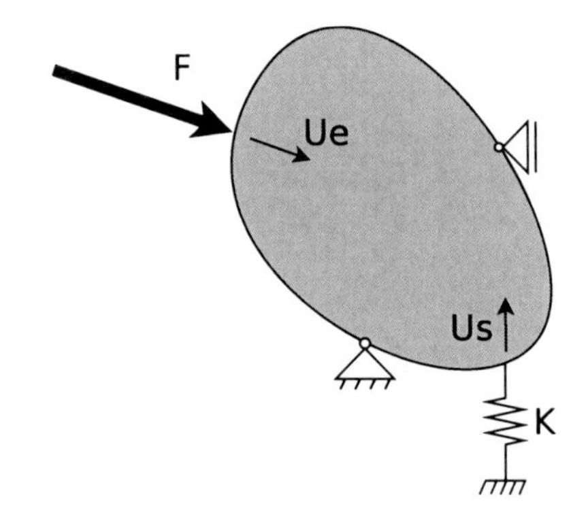

Fig. 1 Initial design domain Ω and the associated boundary conditions

set of constraints to avoid the hinges, the objective is now to maximize the stored elastic energy of the mechanism, leading to a fully compliant mechanisms. The cinematic behavior is imposed by a set of displacement constraints. Unfortunately, the maximization of elastic energy also increases stress levels inside the mechanism (although this issue is not usually addressed in the literature), leading to unrealistic designs. To address this problem, a new constraint is introduced to this formulation, such that the elastic behavior is not violated.

2 Design of Fully Compliant Mechanisms Using Topology Optimization

A compliant mechanism works by transforming some amount of work received through an external agent, W_{in}, into mechanical work delivered to the external medium, W_{out}. Ideally, if the mechanism is composed of rigid parts connected by ideal hinges, all the work received can be delivered and no energy is stored or lost in the process. However, in a flexible mechanism, some energy must be used in order to impose strain (strain energy). As more energy is stored inside the mechanism, more likely it will behave as a real compliant mechanism. Cardoso and Fonseca [3], verified that the existing proposals for the design of compliant mechanism presented in the literature aimed to maximize the energy throughput, thus leading to the appearance of hinges. Those hinges, as pointed by Cardoso and Fonseca, are an artifact used by the optimization process to maximize W_{out}. To obtain a real compliant mechanism, Cardoso and Fonseca proposed the maximization of a function of the stored elastic energy of the mechanism, with the cinematic behavior being imposed by a set of displacement constraints. The formulation is

$$\begin{aligned} &\underset{\mathbf{x}}{Max} \quad \Psi \\ &S.T. \; \mathbf{KU} = \mathbf{F} \\ &\qquad \int_{\Omega} \rho d\Omega \leq V_{max} \\ &\qquad U_j \quad \leq \bar{U}_j \end{aligned} \tag{4}$$

where $\mathbf{x}$ is the vector of design variables, V_{max} constrains the amount of material that can be used to form the part, U_j are components of the displacement vector associated to the kinematic behavior of the mechanism and $\overline{U}_j$ are the imposed values. The main aspect of this formulation is the maximization of Ψ

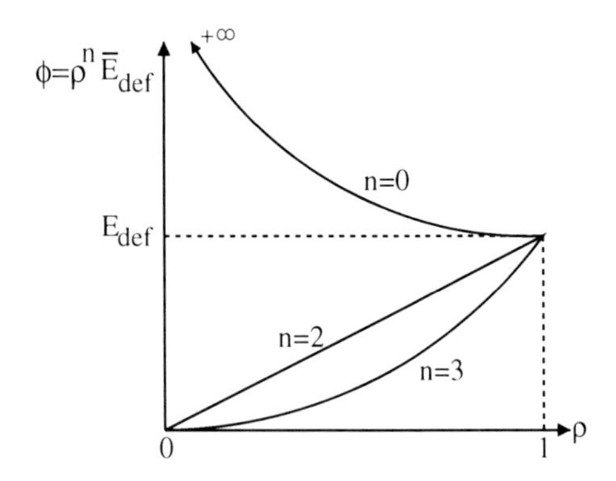

Fig. 2 Modified strain energy with respect to the pseudo density ($p = 1$)

$$\Psi = \sum_{e=1}^{ne} \rho_e^n \mathbf{U}_e^T \mathbf{K}_e \mathbf{U}_e = \sum_{e=1}^{ne} \rho_e^{n+p} \mathbf{U}_e^T \mathbf{K}_e^0 \mathbf{U}_e \tag{5}$$

which can be seen as a modified (penalized) function of the strain energy. This function has a modified asymptotic behavior with respect to the element densities, since the original strain energy of a given element, $\overline{E}_{def} = \mathbf{U}_e^T \mathbf{K}_e \mathbf{U}_e$, becomes infinity as ρ tends to zero [3]. With the modification proposed in Eq. 5, it is possible to adjust the relation between energy and density, by means of the exponent n, as shown in Fig. 2.

In order to obtain some control over the complexity of the topology, we use a vector of intermediate variables, $\mathbf{x}$, located at the nodes of the finite element mesh. These variables are mapped to the element-wise centroidal pseudo densities by a simple spatial average

$$\rho_e = \frac{\sum_{j=1}^{nv} w_{ej} x_j}{\sum_{j=1}^{nv} w_{ej}} \tag{6}$$

where nv is the number of neighbor nodes around element e and w_{ej} are linear weights of the form

$$w_{ej} = \frac{R_{max} - R_{ej}}{R_{max}} \tag{7}$$

where R_{max} is the radius of the filter and R_{ej} is the distance between the coordinates of node j and the centroid of the element e.

This formulation, as presented in [3], is able to generate compliant mechanism without hinges and/or mesh dependency. One main drawback, however, is the fact that stresses are proportional to the strain energy. In order to avoid a non realistic design, as the equilibrium formulation is based on the assumption of elastic behavior, one must impose an additional set of stress constraints.

The correct evaluation of stress constraints imposes many difficulties when using topology optimization. First, as the stress tensor is a local measure, there is a problem associated to the large number of constraints, typically as numerous as the number of elements in the mesh. Despite of the high computational effort, some researchers have successfully adopted this procedure and were able to generate feasible solutions [11]. On the other hand, the use of just one constraint, or a reduced number of them, would be computationally attractive, but one generally has to deal with a weak control of the stress level, as for example, the use of a p-norm constraint of the von Mises stress field. Although it represents a stress field measure, it cannot be related to the maximum stress unless the parameter p tends to infinity, which is numerically infeasible. In this work we use the modified p-norm proposed by Le et al. [10] as stress constraint:

$$c^k \left\| \sigma_{vm}^k \right\|_P \leq \sigma_{lim} \tag{8}$$

with

$$c^k = \frac{\max(\sigma_{vm})^{k-1}}{\left\| \sigma_{vm}^{k-1} \right\|_P} \tag{9}$$

where c^k is a constant dependent of a previous iteration, $k-1$, $\max(\sigma_{vm})^{k-1}$ is the maximum value of the von Mises stress at $k-1$ and σ_{lim} is the limit stress (in this work, we use the yield stress as the limit stress).

This modification relies in the concept that, as k increases and the optimization procedure converges to a given topology, there is a tendency that

$$\frac{\| \sigma_{vm}^k \|_P}{\left\| \sigma_{vm}^{k-1} \right\|_P} \to 1 \tag{10}$$

such that Eq. 8 tends to

$$\max(\sigma_{vm})^{k-1} \leq \sigma lim.$$

A second problem associated to stress constraints in topology optimization is the correct material parametrization. As discussed by Duysinx and Bendsøe [5], to represent the correct stress-strain behavior of a Rank-2 class material, one should use

$$\sigma = \frac{\rho^{\mathbf{p}}}{\rho^{\mathbf{q}}} \varepsilon \tag{11}$$

with $p = q$. In such case, it is observed an undesirable behavior of the stress-strain relation, where even in void regions there should exist stress, making impossible to properly remove material. This leads to ill posed behavior and convergence problems. There are many proposals to circumvent this behavior, such as the ε-relaxation [4] and the use of smooth envelope functions [13]. The singularity associated to the continuous material parametrization can be circumvented by changing the asymptotic behavior of the stress with respect to a change in the material density. The parametrization used in this work is known as qp relaxation [2], where Eq. 11 is used with $q < p$, since it is very ease to use and allows a simple control over the stress behavior.

Thus, the optimization problem of Eq. 4 can now be stated as

$$\begin{aligned} \underset{\mathbf{x}}{Max} \quad & \Psi \\ S.T. \quad & \mathbf{KU} = \mathbf{F} \\ & c^k \left\| \sigma_{vm}^k \right\|_P \leq \sigma_{lim} \\ & \int_\Omega \rho d\Omega \leq V_{max} \\ & U_j \leq \bar{U}_j \end{aligned} \tag{12}$$

The optimization problem is solved using sequential linear programming [8], with a very conservative strategy for the moving limits. If the value of some variable oscillates in the last three iterations, than the moving limit associated to this variable is decreased by a constant factor (0.9) and, if the variable is increasing or decreasing steadily, the moving limit is increased by a factor of 1.1. The maximum move limit allowed is 10 % of the actual value for each variable and the lower bound is 1×10^{-4} %, due to the very non linear nature of the stress constraint.

3 Results

The inverter mechanism with dimensions $100 \times 100 \times 5$ mm, is used to asses the proposed formulation (Fig. 3). The mechanism is subjected to an input force, F and must impose a negative displacement on an external medium (represented by a spring). Due to the vertical symmetry, just half of the domain is modeled, as shown in Fig. 4.

This problem is selected due to its kinematic behavior, resulting in a topology where large amounts of void regions are submitted to relative large strains. As will be shown in this section, this condition is prone to trigger the stress singularity.

The mechanism has four horizontal displacement constraints, U_j, two on the left and two on the right side of the mesh (as labeled in Fig. 5). The distributed load is imposed on the edge of the elements between the left nodes of Fig. 5 and a distributed stiffness is imposed on the edge of the elements between the right nodes of Fig. 5, in order to represent the stiffness of the external medium and to avoid artificial stress singularities associated to point loads and stiffness.

The reference domain, Ω, is discretized using 9,600 four node bilinear isoparametric finite elements. The distributed load is $F_{dist} = 8 \times 10^6 \, \frac{\text{N}}{\text{m}^2}$ and the distributed stiffness is $K_{dist} = 2 \times 10^8 \, \frac{\text{N}}{\text{m}}$. The base isotropic material used in all examples is Copper, with $E^0 = 110$ GPa, $\nu^0 = 0.34$ and $\sigma_{lim} = 60$ MPa. Other parameters

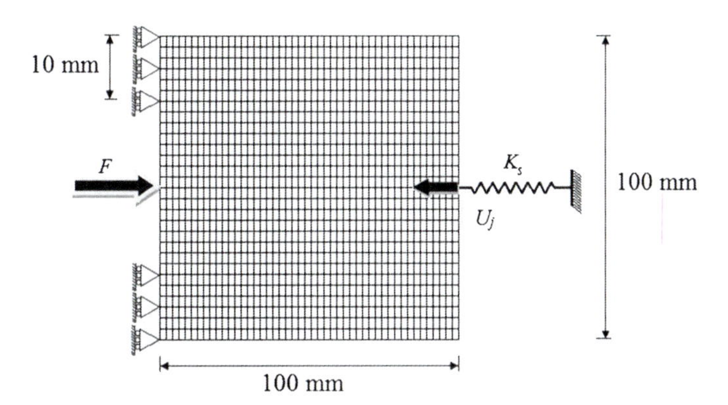

Fig. 3 Inverter mechanism: dimensions and boundary conditions

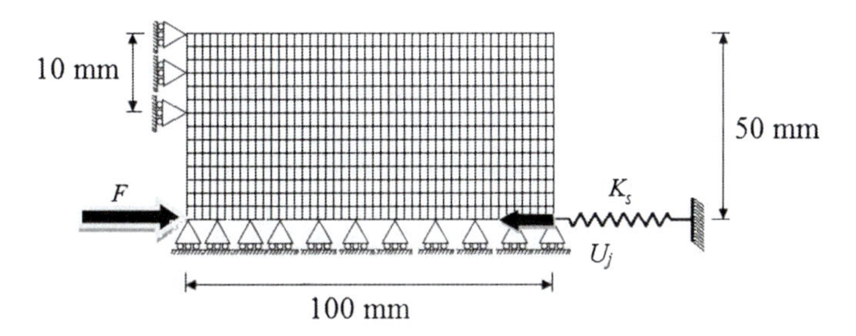

Fig. 4 Model of the inverter mechanism used in this work

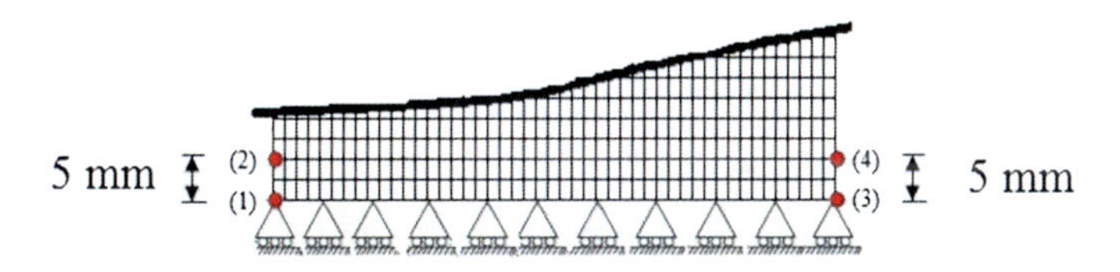

Fig. 5 Nodes used to impose the displacement constraints

used in this section are: $R_{max} = 2 \times 10^{-3}$ m, $\sigma_{lim} = 60$ MPa, $P = 4$, $\mathbf{x}^0 = 0, 5$, $\overline{U}_s = -0, 1$ mm, $\overline{U}_e = 0, 2$ mm, $p = 3$, $q = 2$ and $n = 8$.

It is known that any one of the above parameters can influence the final result. To address the weight of such parameters on the final result, we investigate the influence of the volume constraint, the relaxation parameter q and the influence of the finite element mesh.

Fig. 6 Topologies obtained with an upper limit of 15 % (*left*) and 40 % (*right*) for the volume constraint

3.0.1 Influence of the Volume Constraint

Two volume constraints are used to investigate its influence on the obtained topology: 40 and 15 % of Ω. Figure 6 shows the obtained topologies, where its clear that the amount of material used to form the part has a great influence on the design. Although all the constraints are satisfied, it can be inferred from the results presented in Table 1 that the stress constraint plays a very important role on the design, since for $V_{max} = 40$ just 23 % of base material is used. Also, the input displacement is smaller than the maximum allowed value, since the stress levels are proportional to the input displacements (Fig. 7). So, the stress constraint limits the amount of material used and the magnitude of the imput displacements. When there is no material to spare, $V_{max} = 15$ %, all the allowed material is used (second row in Table 1). This behavior is not presented in the original formulation [3], where no stress constraint was considered.

3.1 Influence of q

The qp parametrization relies on the fact that the stress tensor vanishes when ρ goes to zero, avoiding the singular behavior. This method was proposed by Bruggi [2], where it is shown that for $p < q$ one can obtain a proper behavior for the stress. One very important issue in the design of compliant mechanisms is the fact that some regions of the mesh present large relative elastic deformations. When these regions have low densities elements (void), one can observe a severe singular behavior, even for values of p and q that present a good stress parametrization in structural problems. To show this behavior, the case $V_{max} = 40$ % is chosen with $p = 3$ and $q = 2.8$. As can be seen in Fig. 8, the maximum stresses are distributed in a void area, preventing the topology optimization of achieving convergence. The same problem is than solved with a more relaxed stress parametrization, $q = 2.4$, resulting in a correct stress evaluation inside the domain Ω, as shown in Fig. 9. This behavior is strictly related to the magnitude of strain in void areas, as observed in this kind of mechanism. Table 2 shows the values obtained for both values of q.

Table 1 Results obtained with an upper limit of 15 % (case 1) and 40 % (case 2) for the volume constraint

		Volume fraction (%)	Ψ (J)	Input discpl. (m)	Output displ. (m)	Φ (MPa)
Case 1	$V_{max} = 40$ %	23	0.013	(1) 0.000158, (2) 0.000152	(3) −0.000102, (4) −0.000100	60
Case 2	$V_{max} = 25$ %	23	0.00625	(1) 0.000107, (2) 0.000105	(3) −0.000102, (4) −0.000100	60
Case 3	$V_{max} = 15$ %	15	0.0047	(1) 0.000114, (2) 0.000113	(3) −0.000106, (4) −0.000100	60

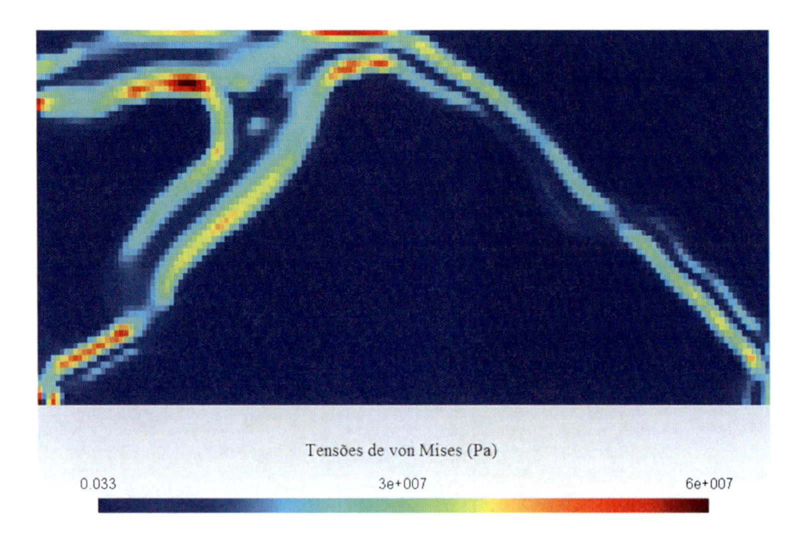

Fig. 7 Distribution of the von Mises equivalent stress for $V_{lim} = 40\ \%$

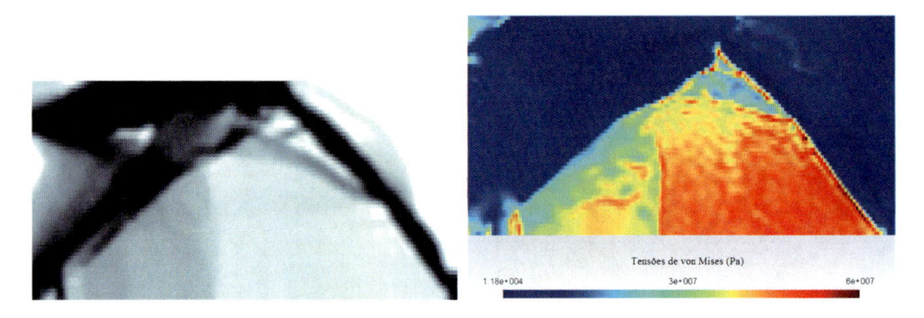

Fig. 8 Topology and distribution of the equivalent von Mises stress for $V_{lim} = 40\ \%$, $p = 3$ and $q = 2.8$

3.2 *Influence of the Finite Element Mesh*

The number of elements in the finite element mesh has a direct consequence on the quality of displacements and stresses. Also, when dealing with Topology Optimization, a finer mesh also allows a finer description of the boundary.

Two meshes where used to assess the influence of the mesh refinement on the design. The first mesh has 9,600 elements, as in the previous examples, while the second mesh has 38,400 finite elements. One interesting result of mesh refinement is the fact that for 9,600 elements a proper stress relaxation is obtained with $q = 2.0$ but with 38,400 finite elements the maximum equivalent stress occurs in elements with intermediate density, as can be seen on the left of Fig. (10). To avoid such behavior, the same case was solved with $q = 1.5$, resulting in a clear topology (11).

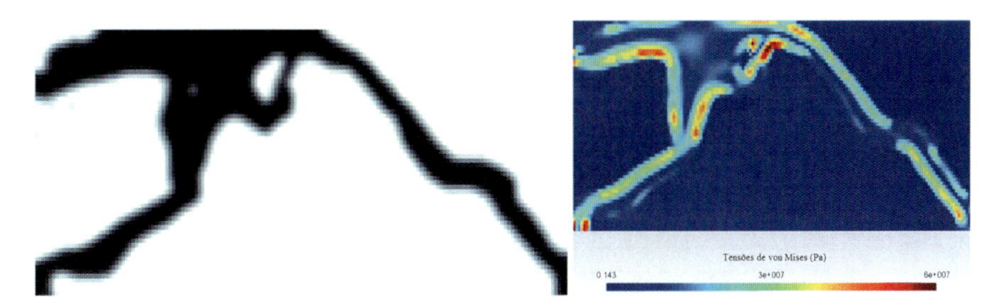

Fig. 9 Topology and distribution of the equivalent stress for $V_{lim} = 40\ \%$, $p = 3$ and $q = 2.4$

Table 2 Results obtained for different values of q

q	Volume fraction (%)	Ψ (J)	Input displ. (m)	Output displ. (m)	Φ (MPa)
$V_{max} = 40\ \%$ 2, 840		0.0008	(1) 0.000013, (2) 0.000012	(3) −0.000010, (4) −0.000010	60
$V_{max} = 40\ \%$ 2, 423		0.008	(1) 0.000120, (2) 0.000110	(3) −0.000118, (4) −0.000100	60

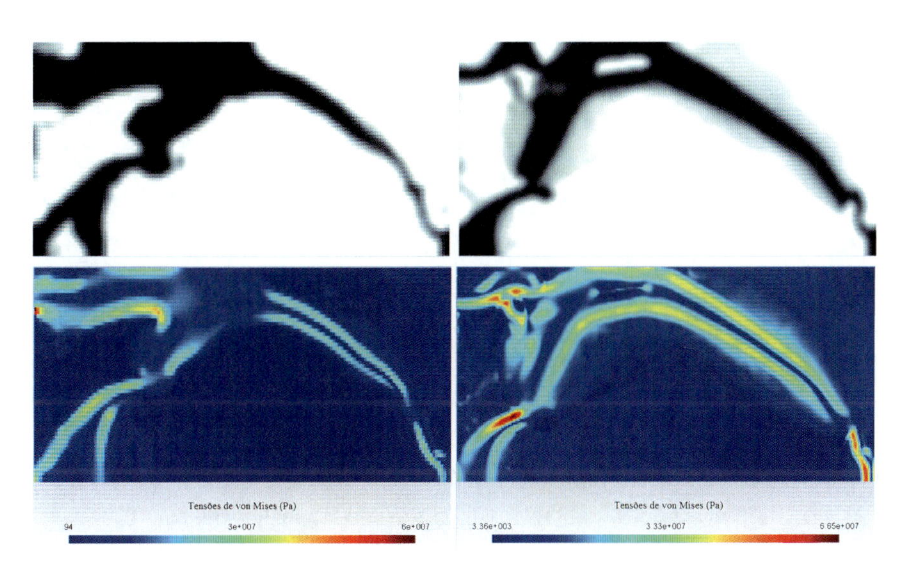

Fig. 10 Topology and distribution of the equivalent von Mises stress for 9,600 finite elements (*left*) and 38,400 finite elements (*right*), $q = 2$

This behavior is not addressed in the literature, since the evaluation of the quality of the mesh is rarely discussed in topology optimization articles. Also, as the bilinear isoparametric finite element is very stiff in bending, the mesh refinement has a direct consequence on the amount of strain in the void region, explaining the low values needed for q in order to avoid the singularity. Table 3 shows the results obtained for both values of q.

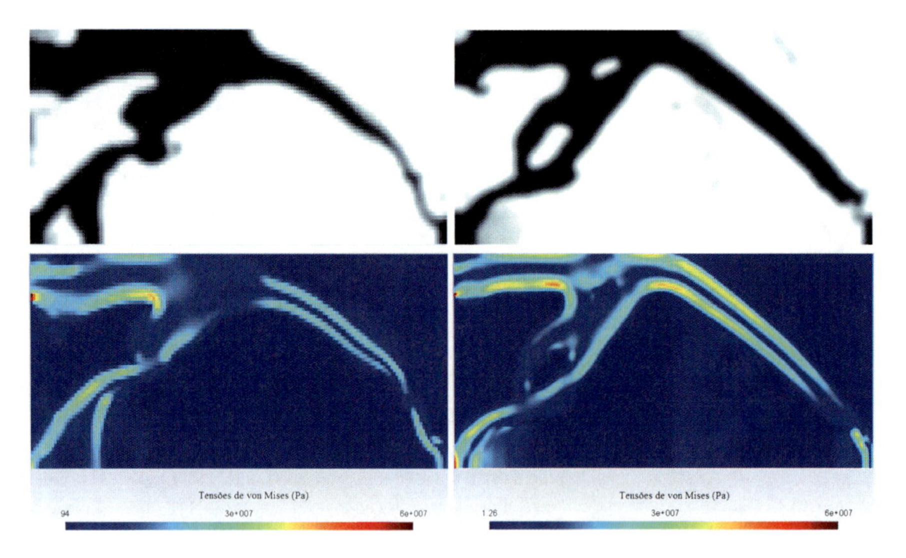

Fig. 11 Topology and distribution of the equivalent von Mises stress for 9,600 finite elements (*left*) and 38,400 finite elements (*right*), $q = 1.5$

Table 3 Results obtained with the refined finite element discretization

	q	Final volume fractionl (%)	Ψ (J)	Input displ. (m)	Output displ. (m)	Φ (MPa)
$ne = 38400$	2, 0	24	0.01022	(1) 0.000138, (2) 0.000137	(3) −0.000127, (4) −0.000121	66.5
$ne = 38400$	1, 5	28	0.0121	(1) 0.000110, (2) 0.000110	(3) −0.000110, (4) −0.000110	60

4 Concluding Remarks

The results obtained in the previous section attest that the proposed formulation is able to produce a feasible compliant mechanism with a clear and free-hinge topology.

Analyzing the results, it is clear that the stress constraint has a deep impact in the original formulation proposed by Cardoso and Fonseca [3], as the maximization of the strain energy leads to a maximization of some displacements and, consequently, maximum stresses. Thus, to satisfy the stress constraint, it is sometimes necessary to decrease the displacements. As the formulation is based in energy, decreasing the work delivered to the mechanism also decreases the energy used to form the compliant part. This can create a conflict between the displacement and stress constraints, as a severe imposition in one of them can result in an infeasible design. A severe stress constraint can hinder the storage of strain-energy and consequently will hinder the fundamental requirement of the proposed formulation: the design of fully compliant mechanisms.

Moreover, the volume constraint remained inactive in some cases, as the displacement constraints exert a control over final volume. Likewise, as a stress constraint has a direct effect on the magnitude of the displacements, it can be stated that the stress constraint also impacts on volume constraints.

The qp-relaxation was effective to control the singularity phenomenon, penalizing the stresses in low density elements when $q < p$. One of the main conclusions is that appropriate parameters for the qp-relaxation in Topology Optimization for mechanical structures do not necessarily lead to appropriate parameters in the design of compliant mechanisms, as the strain field associated to the imposition of the kinematic behavior can increase the amount of elastic strain in some regions of the finite element mesh.

Another important result was obtained with the finite element mesh refinement, since an improvement in the evaluation of the stress field also affects the choice of the q parameter. Thus, it must be emphasized that an extreme care must be taken regarding the choice q parameter of qp-relaxation, since it depends on the relative magnitude of the strain field, which is related to the kinematic behavior of the part and also to the quality of the stress evaluation.

It can be observed that the topologies obtained in this work have a different shape when compared with other formulations used to design compliant mechanisms. It can be explained by the fact that the topologies obtained have a more "curved" shape in order to store more elastic energy when subjected to bending in order to maximize the objective function.

Acknowledgments All the post-processing images were generated using the free and general pre and post-processing program *gmsh* [7]. The linear programming problem was solved using the free SLATEC package *dsplp* [8].

References

1. Bendsøe, M.P., Sigmund, O.: Topology Optimization: Theory. Springer, New York (2003)
2. Bruggi, M.: On an alternative approach to stress constraints relaxation in topology optimization. Struct. Multidis. Optim. **36**, 125–141 (2008)
3. Cardoso, E.L., Fonseca, J.S.: Strain energy maximization approach to the design of fully compliantmechanisms using topology optimization. Latin Am. J. Solids Struct. **1**, 263–275 (2004)
4. Cheng, G.D., Guo, X.: ϵ-relaxed approach in structural topology optimization. Struct. Optim. **13**, 258–266 (1997)
5. Duysinx, P., Bendsøe, M.P.: Topology optimization of continuum structures with local stress constraints. Int. J. Numer. Methods Eng. **43**, 1453–1478 (1998)
6. Frecker, M.I., Kikuchi, N., Kota, S.: Optimal synthesis of compliant mechanisms to satisfy kinematic and structural requirements—preliminary results. In: Proceedings of the 1996 ASME Design Engineering Technical Conferences and Computers in, Engineering Conference, pp. 177–192 (1996)
7. Geuzaine, C., Remacle, J.F.: Gmsh: a three-dimensional finite element mesh generator with built-in pre- and post-processing facilities. Int. J. Numer. Meth. Eng. **79**(11), 1309–1331 (2009)
8. Hanson, R., Hirbert, K.: A sparse linear programming subprogram. Tech. rep., Sandia National Laboratories (1981). SAND81-0297

9. Kikuchi, N., Nishiwaki, S., Fonseca, J.S.O., Silva, E.C.N.: Design optimization method for compliant mechanisms and material microstructure. Comput. Methods Appl. Mech. Eng. **151**, 401–417 (1998)
10. Le, C., Norato, J., Bruns, T., Ha, C., Tortorelli, D.: Stress-based topology optimization for continua. Struct. Multidis. Optim. **41**, 605–620 (2010)
11. Pereira, J.T., Fancello, E.A., Barcellos, C.S.: Topology optimization of continuum structures with material failure constraints. Struct. Multidis. Optim. **26**, 50–56 (2004)
12. Poulsen, T.A.: A simple scheme to prevent checkerboard patterns and one-node connected hinges in topology optimization. Struct. Multidis. Optim. **24**, 396–399 (2002)
13. Rozvany, G.I.N., Sobieszczanski-Sobieski, J.: New optimality criteria methods: forcing uniqueness of the adjoint strains by cornerrounding at constraint intersections. Struct. Multidis. Optim. **4**, 244–246 (1992)
14. Sigmund, O.: On the design of compliant mechanisms using topology optimization. Mech. Struct. Mach. **25**, 495–526 (1997)
15. Wang, F., Lazarov, B.S., Sigmund, O.: On projection methods, convergence and robust formulations in topology optimization. Struct. Multidis. Optim. **43**, 767–784 (2011)

A Genetic Algorithm for Optimization of Hybrid Laminated Composite Plates

M. A. Luersen and R. H. Lopez

Abstract This chapter presents a genetic algorithm (GA) to pursue the optimization of hybrid laminated composite plates. Fiber orientation (predefined ply angles), material (glass-epoxy or carbon-epoxy layer) and total number of plies are considered as design variables. The constraints of the optimization problem are taken into account by a multiplicative dynamic penalty approach. The GA is chosen as an optimization tool because of its ability to deal with non-convex, multimodal and discrete optimization problems, of which the design of laminated composites is an example. First, the developed algorithm is detailed explained and, in the first example, validated by comparing its results to other obtained from the literature for non-hybrid laminates. Then, two examples of material cost minimization of hybrid laminates are solved, under the constraints of a maximum weight and buckling or ply failure. In the example where the ply failure is used as a constraint, three different criteria are tested independently: maximum stress, Tsai-Wu and the Puck failure criterion and the results yielded by them are compared and discussed. It was found that each failure criterion yielded a different optimal design.

1 Introduction

The laminated composite materials commonly used in the design of high-performance structures—and considered in this study—are made by stacks of layers, each layer usually composed by a matrix of polymeric material reinforced by continuous fibers

M. A. Luersen(✉)
Department of Mechanical Engineering, Federal University of Technology– Paraná (UTFPR), Av. Sete de Setembro, 3165, Curitiba, PR80230-901, Brazil
e-mail: luersen@utfpr.edu.br

R. H. Lopez
Department of Civil Engineering, Federal University of Santa Catarina (UFSC), Rua João Pio Duarte Silva, s/n, Florianópolis, SC88040-900 , Brazil
e-mail: rafaelholdorf@gmail.com

P. A. Muñoz-Rojas (ed.), *Optimization of Structures and Components*, Advanced Structured Materials 43, DOI: 10.1007/978-3-319-00717-5_4,

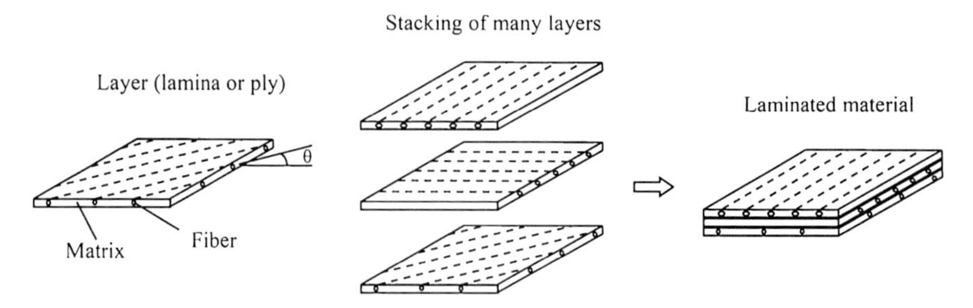

Fig. 1 Laminated composite material

oriented in a specific direction. They give the designer the possibility to tailor the material according to the application and the structures formed by these materials present high stiffness/mass and strength/mass ratios. Figure 1 shows a scheme of a laminated composite material formed by unidirectional layers.

Laminated composites can provide not only high structural performance, but also reduce drastically the direct and/or operational costs of such structures. A recent report by the US National Materials Advisory Board estimates that a 1 lb weight reduction amounts to a total saving of $200 over the 100,000 h life of a civil transport, increasing to $1,000 in the case of military aircraft and reaching $20,000 for spacecrafts [1]. These numbers put in evidence the importance of the structural weight saving in aerospace industry. Hence, to achieve the best results, optimization techniques have been developed and employed in the design of laminated composite structures. Among these techniques, the genetic algorithm (GA) has been widely used to determine the optimal design of composite structures [2–7]. It has passed over 20 years since the pioneer works of Callahan and Weeks [8] and Le Riche and Haftka [3] and the GA is still subject of current research in this area.

The GA is highly suited for the optimization of laminated composites especially because:

- it is able to handle discrete design variables;
- it does not require gradient information and can be applied to problems where the gradient is hard to obtain or simply does not exist;
- if correctly tuned, it does not get stuck in local minima;
- it can be applied to non-smooth or discontinuous functions; and
- it furnishes a set of optimal solutions instead of a single one, thus giving the designer a set of options.

On the other hand, it is subject of current research due to its known drawbacks, which include the following:

- it requires the tuning of many parameters by trial and error to maximize efficiency;
- *a priori* estimation of its performance is an open mathematical problem; and
- a large number of evaluations of the objective function are required to achieve optimization, which can make the use of GA nonviable depending on the computational cost of each evaluation.

In the design of laminated composites, the ply thicknesses are often predetermined and the ply orientations are usually restricted to a small set of angles due to manufacturing limitations and/or limited availability of experimental data. For example, usually only 0, 45 and 90 degree ply orientations are available for manufacturing. This leads to problems of discrete or stacking-sequence optimization.

Many objective functions have been introduced, such as the buckling load [3, 6] (to be maximized), the stiffness in one direction (to be maximized), and the strength [9] (to be maximized), as well as the cost or weight [4, 10–12] (to be minimized). The GA has also solved optimization problem in which the frequency response of the structure was included in the objective function, e.g. [13–15], to name just a few, included the maximization of fundamental frequencies in their optimization problems. However, in engineering design, often we need to optimize several objective functions simultaneously, e.g. minimize the cost and maximize the performance. Hence, multi-objective optimization techniques based on GA have also been applied to the design of composite structures. See, for instance, references [16–21].

In addition to the failure criterion, other restrictions are usually involved in the optimal design of laminated composites. Examples of such restrictions in the literature include upper and bound limits for the design variables, laminate symmetry and balance, and a maximum number of contiguous plies with the same orientation angle. The latter restriction is often used to prevent matrix cracking. Liu et al. [6] applied and compared repair strategies using a maximum number of contiguous plies with the same orientation angle and showed that the Baldwinian repair strategy drastically reduces the computational cost of constrained optimization (the cost of the GA, measured by the number of function evaluations necessary to achieve the optimization, is reduced by one or two orders of magnitude). The first ply failure constraint is often handled by a penalty approach [3, 4].

When the optimization problem involves continuous and discrete variables (a mixed-variable problem), the representation of the variables in a single string in the GA increases the dimension of the design variables space (since one real variable is transformed into many integer ones). To overcome this problem, decomposition approaches have been introduced [22]. These often involve decomposition at local and global levels: at the local level the best stacking sequence is determined for a given geometry, and at the global level new geometries are generated based on the result furnished by the local level. Such an approach was adapted for optimal composite design by Murugan et al. [17]. Antonio [23] introduced a hierarchical GA with age structure adapted for the optimal design of hybrid composite structures with multiple solutions. The algorithm he proposed addressed the optimal stacking sequence and material topology as a multimodal optimization problem. Antonio showed that the procedure for species control is effective because it allows multiple optimal solutions and guarantees subpopulation diversity.

With regard to failure criteria, one of the main criticisms of many studies related to optimal composite design is the use of von Mises or Hill yield based criteria [e.g. maximum stress (MS), Tsai-Wu (TS)], which are more suitable for ductile materials [24]. In fact, as the failure behavior of composite parts is similar to that of brittle material, it would be more appropriate to use criteria suited to materials that exhibit brittle fractures, such as Mohr's criterion. A suitable criterion for composites that takes this fracture behavior into account is, for instance, the Puck failure criterion (PFC) [24, 25].

Based on the previous works of the authors [7, 26], this study presents a genetic algorithm to pursue the optimization of hybrid laminated composite structures. Fiber orientation (predefined ply angles), material (glass-epoxy or carbon-epoxy layer) and total number of plies are considered as design variables. First, the developed algorithm is validated by comparing their results to other found in the literature [27]. Then, three different failure criteria, the MS, TW and the PFC, are used as constraint in the optimization and their results are compared. The chapter is organized as follows. In Section Genetic Algorithm the chromosome representation, the genetic operators and the constraint handling are described. Section Failure Criteria presents a brief review of the failure criteria used in this work. The problem statements and numerical results are shown in Section Numerical Results. Finally, Section Conclusions reports the main conclusions that were drawn.

2 Genetic Algorithm

Genetic algorithms loosely parallel biological evolution and were originally inspired by Darwin's theory of natural selection. The specific mechanics of genetic algorithms often use the language of microbiology, and their implementation frequently mimics genetic operations [28]. A GA generally involves genetic operators (such as crossover and mutation) and selection operators intended to improve an initial random population. Selection usually involves a fitness function characterizing the quality of an individual in terms of the objective function and the other elements of the actual populations. Thus, a GA usually starts with the generation of a random initial population and iterates by generating a sequence of populations from the initial one. At each step the genetic operators are applied to generate new individuals. The fitness of each available individual is computed and the whole population is ranked according to increasing fitness. A subpopulation is then selected to form a new population. Many selection methods may be found in the literature. In this work, tournament selection is applied (see reference [29]). At this point the algorithm may repeat the process or, before that, a local search may be pursued, which is the case of this approach. Two local search methods are employed, named as: Neighborhood Search and Material Grouping [27]. Then, all the procedure is repeated until a stopping condition is satisfied.

The genetic operators employed in this work are crossover, mutation, gene swap, stack-deletion and stack-addition. In the following the constraint handling, chromosome coding and the genetic operators are shown in detail.

2.1 Constraint Handling

In GAs, the most common ways of handling constraints are data structure, repair strategies and penalty functions [30]. The symmetry and balance of the laminate are handled by using the data structure strategy, which consists of coding only half of the laminate and considering that each stack of the laminate is formed by two laminae with the same orientation but opposite signs (for instance, $\pm 45°$). A double-multiplicative dynamic penalty approach [30] is used here to take into account the other constraints (i.e., failure criteria, buckling, weight limit). This approach leads to a penalty term being added to the objective function: this supplementary term has a multiplicative form and involves autonormalization. It is written as

$$P(\mathbf{x}, q) = \prod_{j=1}^{m} \left[1 + \frac{\hat{g}_j(\mathbf{x})}{b_j} \left(\frac{q+Q}{Q} \right) \right] , \tag{1}$$

where m is the number of constraints, x is the vector of the design variables, $\hat{g}_j(\mathbf{x})$ is the constraint violation, b_j is a normalization parameter, q is the current generation number and Q is the total number of generations. The main advantage of this approach is that the penalization parameters do not need to be tuned.

In the first numerical example (see Sect. 4.1) there is a constraint over the number of contiguous plies with the same orientation, which is dealt by the repair strategy developed by Todoroki and Haftka [5]. This strategy consists in scanning the chromosome—from the outmost to the innermost plies—and repairing the each gene that does not fulfill the contiguity constraint.

2.2 Chromosome Representation

The classical binary representation is not used here; instead, the allowed angle values represent the genes of the chromosomes (i.e., $[0_2 \ldots \pm 45 \quad 90_2]$). In our numerical examples, we consider the optimization of a hybrid laminated composite. In this case, each individual in the population is represented by two chromosomes: the first describes the angle of orientation of the layers, and the second the layers materials. Table 1 details this approach.

The orientation angle and material of each ply are coded in the chromosomes of each individual. One example of chromosome decoding is shown in Fig. 2.

2.3 Crossover

Crossover is the basic genetic operator. It involves combining the information from two parents to create one or two new individuals. The X1-thin crossover operator

Table 1 Chromosome representation

Stack orientation chromosome (C1)			
Empty stack	0_2	±45	90_2
0	1	2	3
Stack material chromosome (C2)			
Empty stack	Carbon-epoxy (CE)	Glass-epoxy (GE)	
0	1	2	

$$\text{C1: } [0\ \ 2\ \ 1\ \ 3\ \ 2] \longrightarrow [\pm45\ \ 0_2\ \ 90_2\ \ \pm45]_S$$
$$\text{C2: } [0\ \ 1\ \ 2\ \ 2\ \ 1] \longrightarrow [\text{CE}\ \ \text{GE}\ \ \text{GE}\ \ \text{CE}]_S$$

Fig. 2 Example of chromosome decoding

		Before	After
Parent 1	C1	0 0 3 2 1 ⋮ 1 3	0 0 3 2 1 ⋮ 2 1
	C2	0 0 1 1 2 ⋮ 2 2	0 0 1 1 2 ⋮ 1 2
Parent 2	C1	0 0 0 1 2 ⋮ 2 1	0 0 0 1 2 ⋮ 1 3
	C2	0 0 0 1 1 ⋮ 1 2	0 0 0 1 1 ⋮ 2 2

Fig. 3 Example of crossover operation

is presented in Le Riche and Haftka [3] and used here. In Le Riche and Haftka [3] several crossover methods are described and compared. The X1-thin is a one-point crossover strategy that restricts the location of the breakpoint to the full part of the thinner parent laminate and generates two new individuals, as shown in Fig. 3.

2.4 Mutation

The mutation operator must be applied in the GA to guarantee gene diversity so that the algorithm does not get stuck in local minima [29]. In the present work, this operator is based on randomly changing the value of a gene in the chromosome. Thus, the algorithm chooses an individual of the population and then, also randomly chooses a gene to be mutated. As the work is dealing with hybrid laminated composites, the mutation is applied to the angle as well as to the material chromosome. Figure 4 exemplifies both mutations.

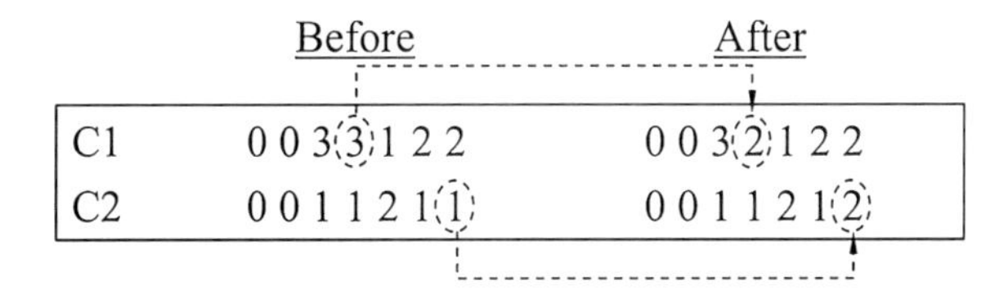

Fig. 4 Example of mutation operator

	Before	After
C1	0 0 3 3 1 2 2	0 0 3 2 1 2 3
C2	0 0 1 1 2 1 2	0 0 1 2 2 1 1

Fig. 5 Example of gene-swap operator

2.5 Gene-Swap

The gene-swap operator selects two genes randomly from a laminate and then swaps them. It was introduced by Le Riche and Haftka [11], who showed that the permutation operator [11], which was used before the gene-swap operator was introduced, shuffles the digits too much and that the gene-swap operator is more efficient. The same authors showed that this operator is quite effective when the problem deals with buckling load, because when an individual has a good pool of genes, the gene-swap helps to reorganize them, possibly resulting in a better design than when the mutation is applied. One example of gene-swap for hybrid laminate chromosome is given in Fig. 5.

In the example of the hybrid laminated composite referred to above, the crossover points of the two chromosomes for each individual are the same. In addition, when the gene-swap is applied, both orientation and material are swapped.

2.6 Stack Addition-Deletion

We introduce two supplementary operators. The first one adds and the second deletes a lamina of the composite part under design. The first operator tends to force the laminate to satisfy the first ply failure constraint, while the second one tends to reduce the weight of the laminate, thus forcing the laminate to satisfy the criterion of minimum weight. An example of this operator is in Fig. 6.

Both operators always act on the lamina closest to the mid-surface of the laminate, since it has the weakest effect on the bending properties of the structure. This feature may be important when buckling is involved, since buckling is highly dependent on the bending properties of the laminate. It is more convenient to delete the lamina

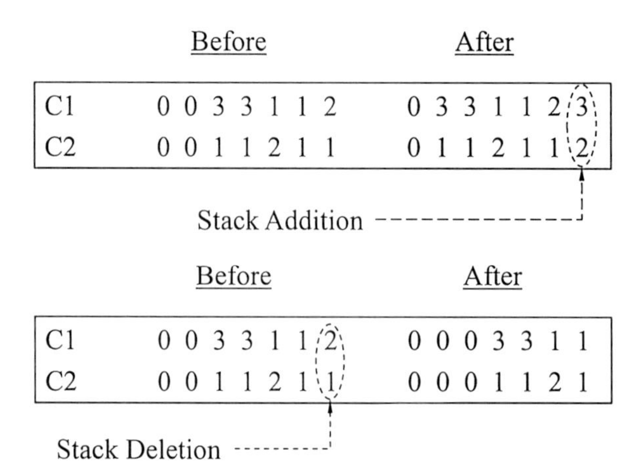

Fig. 6 Example of stack addition-deletion operators

with the weakest influence on the bending properties, since it is observed in practice that the algorithm rapidly converges to the best design for the most external laminae.

2.7 Local Search

Local search may be employed in the GA in order to accelerate the convergence of the algorithm. As already mentioned, two local search methods are used: Neighborhood Search and Material Grouping.

The Neighborhood Search is based on making small changes in the genes of the chromosome that represents the orientations (C1). One individual is chosen among the best designs of the population (i.e., among the five best individuals). Then, one of its genes is randomly chosen and the other allowable values that this gene may have are tested. Finally, the best design is chosen to remain in the population (i.e., the best design among individual, test 1 and test 2, as shown in Fig. 7).

In the practice of optimizing laminated composites, it has been noted that usually the different material laminas are grouped together at the optimal design [27]. Thus, the Material Grouping method consists in grouping the material laminas of a

	C1	C2
Individual	0 0 3 3 1 2 2	0 0 1 1 2 1 2
Test 1	0 0 1 3 1 2 2	0 0 1 1 2 1 2
Test 2	0 0 2 3 1 2 2	0 0 1 1 2 1 2

Fig. 7 Example of Neighborhood Search operator

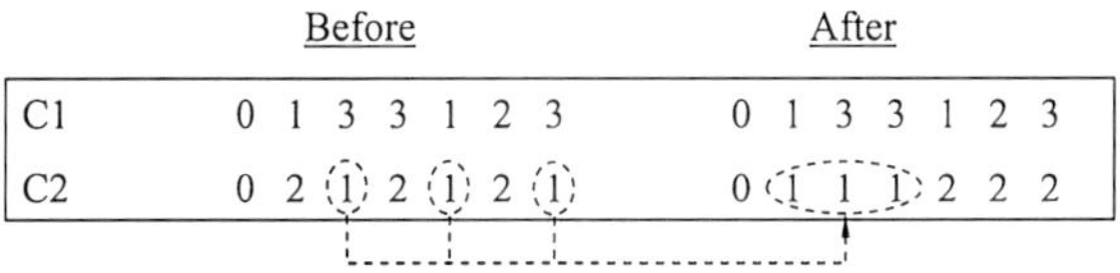

	Before	After
C1	0 1 3 3 1 2 3	0 1 3 3 1 2 3
C2	0 2 1 2 1 2 1	0 1 1 1 2 2 2

Fig. 8 Example of Material Grouping operator

chromosome. First, one individual is chosen among the best ones of the population. Then, the genes with the same material are grouped and the new design is tested. If such design is better than the original one, it is kept in the population. The Material Grouping operation is depicted in Fig. 8.

3 Failure Criteria

In order to keep the manuscript self-contained, in this section we summarize the main features of the composite failure models used in the numerical examples.

Failure analysis of laminated composites is usually based on the stresses in each lamina in the principal material coordinates [31] (see Fig. 9). The failure criteria can be classified in three classes: limit or non-interactive theories (e.g., maximum stress or maximum strain), interactive theories (e.g., Tsai-Hill, Tsai-Wu or Hoffman) and partially interactive or failure mode-based theories (PFC) [32]. In Example 3 (see Sect. 4.3), one criterion from each class is considered. Thus, they are described in the following.

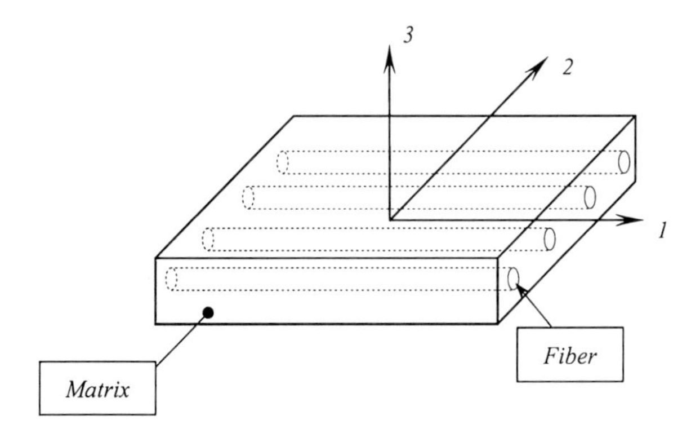

Fig. 9 Principal material coordinate system of a unidirectional lamina

3.1 Maximum Stress Failure Criterion

According to the maximum stress theory, failure occurs in a layer when a maximum stress in the principal material coordinates exceeds the respective strength. That is,

$$\begin{array}{ll} \sigma_1 \geq X_T \text{ or } \sigma_2 \geq Y_T & \text{(for tensile stresses)} \\ \sigma_1 \leq -X_C \text{ or } \sigma_2 \leq -Y_C & \text{(for compressive stresses)} \\ |\tau_{12}| \geq S_{12} & \text{(for shearing stresses)} \end{array} \tag{2}$$

where σ_1 and σ_2 are the normal stresses in the directions 1 and 2, respectively; τ_{12} is the shear stress in the elastic symmetry plane 1–2; X_T and X_C are the tensile and compressive strengths parallel to the fiber direction, respectively; Y_T and Y_C are the tensile and compressive strengths normal to the fiber direction, respectively; and S_{12} is the shear strength. Note that X_T, X_C, Y_T, Y_C and S_{12} are positive quantities.

3.2 Tsai-Wu Failure Criterion

The Tsai-Wu criterion, formulated to predict failure of orthotropic materials, is derived from the von Mises yield criterion. It states that the lamina fails when the following condition is satisfied

$$F_{11}\sigma_1^2 + 2F_{12}\sigma_1\sigma_2 + F_{22}\sigma_2^2 + F_{21}\tau_{12}^2 + F_1\sigma_1 + F_2\sigma_2 \geq 1 \tag{3}$$

where F_{ij} are parameters that are a function of the strength properties X_T, X_C, Y_T, Y_C and S_{12} (see for instance, reference [31]).

3.3 Puck Failure Criterion

In this section, only the main features of the PFC are presented. The entire derivation can be found in references [24, 25]. The PFC follows Mohr's hypothesis that fracture is caused exclusively by the stresses that act on the fracture plane. It involves two main failure modes: Fiber Failure (FF) and Inter-Fiber Failure (IFF) [24].

FF is based on the assumption that fiber failure under multiaxial stresses occurs at the same threshold level at which failure occurs for uniaxial stresses. After some development, Puck and Schürmann [24] found that failure occurs if one of the following conditions is satisfied:

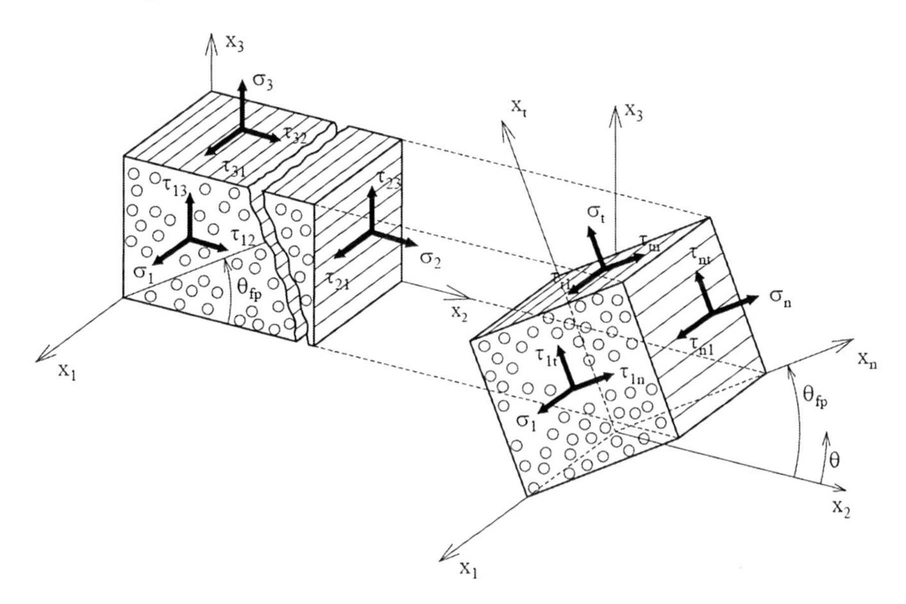

Fig. 10 Transformation from the 1-2-3 axes to the axes corresponding to the failure plane (σ_n, τ_{nt}, τ_{n1})

$$\frac{S}{\varepsilon_{1T}} = 1\, if\ S \geq 0; \quad -\frac{S}{\varepsilon_{1C}} + (10\gamma_{21})^2 = 1 \quad \text{otherwise;}$$
$$\text{where } S = \left(\varepsilon_1 + \frac{\upsilon_{f12}}{E_{f1}} m_{\sigma 1} \sigma_2\right) \tag{4}$$

where ε_{1T} and ε_{1C} are tensile and compressive failure strains in direction 1, respectively; e1 is the normal strain in the direction 1; ν_{f12} is the Poisson's ratio of the fibers (the ratio of the strain in direction 2 to the strain in direction 1, both of which are caused by a stress in direction 1 only); $m_{\sigma f}$ accounts for a stress-magnification effect caused by the different moduli of the fibers and matrix (in direction 2), which leads to an uneven distribution of the stress σ_2 from a micromechanical point of view [24]; E_{f1} is the Young's modulus of the fiber in direction 1; and γ_{21} is the shear strain in the plane 1-2. Note that S $\geq$ 0 corresponds to tension, while S $<$ 0 corresponds to compression.

Instead of dealing with the principal material coordinates (axes 1-2-3), IFF equations are derived based on the axes corresponding to the failure plane. These axes are shown in Fig. 10, where θ_{fp} represents the angle at which failure occurs. The PFC therefore provides not only a failure factor, but also the inclination of the plane where failure will probably take place, thus allowing a much better assessment of the consequences of IFF in the laminate.

IFF is subdivided into three failure modes, as described in reference [24], which are referred to as A, B and C. These are shown in Fig. 11. Mode A occurs when the lamina is subjected to tensile transverse stress, whereas modes B and C correspond

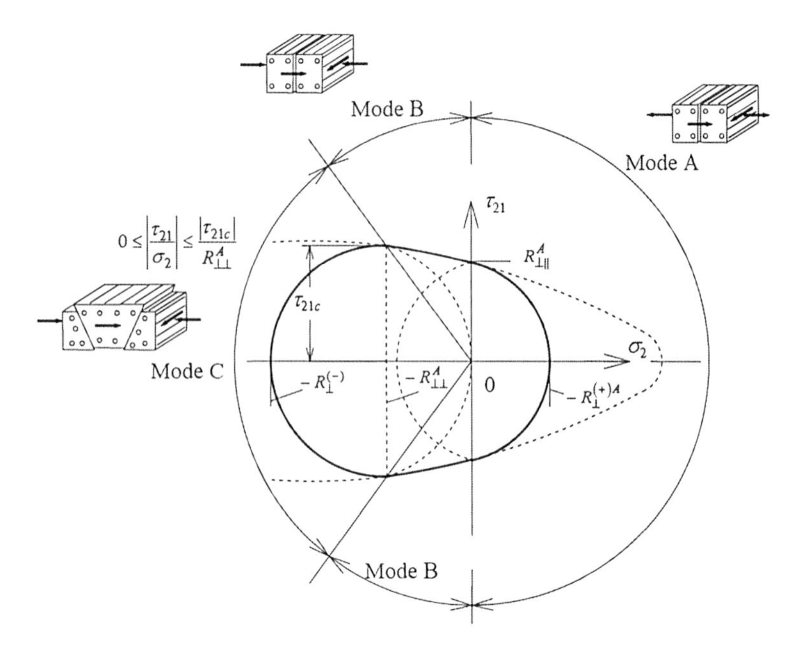

Fig. 11 (σ_2, τ_{21}) fracture curve for σ_1, representing the three different fracture modes (A, B and C) for the PFC

to compressive transverse stress. The classification is based on the idea that a tensile stress $\sigma_n > 0$ promotes fracture, while a compressive stress $\sigma_n < 0$ impedes shear fracture. For $\sigma_n < 0$, the shear stresses τ_{nt} and τ_{n1} (or just one of them) have to face an additional fracture resistance, which increases with $|\sigma_n|$, analogously to an internal friction [24]. The distinction between modes B and C is based on their failure angles, which are 0° for mode B and a different value for mode C. In addition, failure mode C is considered more severe, since it produces oblique cracks and may lead to serious delamination.

The equations for the PFC are summarized in Table 2, where we also introduce a weakening factor f_w, which decreases the strength of the laminate due to high stress in the fiber direction. According to Puck and Schürmann [24], f_w is given by

$$f_w = \left(0.9 f_{E(FF)}\right)^n \tag{5}$$

where $f_{E(FF)}$ is the failure effort for FF in the lamina, and n, in this equation, is an exponent that depends on the matrix of the laminate (for instance, $n = 6$ for epoxy). We refer henceforth to this situation as PFC_fw, while we denote the situation where $f_w = 0$ by PFC.

Table 2 Equations for the PFC [24]

Type of failure	Failure mode	Failure condition ($f_{E(FF)} or f_{E(IFF)}$)	Condition for validity
Fiber failure (FF)	Tensile	$\frac{S}{\varepsilon_{1T}} = 1$	if $S \geq 0$
$S = \varepsilon_1 + \frac{\upsilon_{f12}}{E_{f1}} m_{\sigma 1} \sigma_2$	Compressive	$-\frac{S}{\varepsilon_{1C}} + (10\gamma_{21})^2 = 1$	if $S < 0$
Inter fiber failure (IFF)	Mode A	$\sqrt{\left(\frac{\tau_{21}}{S_{21}}\right)^2 + \left(1 - p_{\perp\|\|}^{(+)} \frac{Y_T}{S_{21}}\right)^2 \left(\frac{\sigma_2}{Y_T}\right)^2} + p_{\perp\|\|}^{(+)} \frac{\sigma_2}{S_{21}} + f_w = 1$	$\sigma_2 \geq 0$
	Mode B	$\frac{1}{S_{21}} \left(\sqrt{\tau_{21}^2 + \left(p_{\perp\|\|}^{(-)} \sigma_2\right)^2} + p_{\perp\|\|}^{(-)} \sigma_2 \right) + f_w = 1$	$\sigma_2 < 0$ and $0 \leq \left\|\frac{\sigma_2}{\tau_{21}}\right\| \leq \frac{R_{\perp\perp}^A}{\|\tau_{21c}\|}$
	Mode C	$\left[\left(\frac{\tau_{21}}{2\left(1 + p_{\perp\perp}^{(-)} S_{21}\right)} \right)^2 + \left(\frac{\sigma_2}{Y_C}\right)^2 \right] \frac{Y_C}{(-\sigma_2)} + f_w = 1$	$\sigma_2 < 0$ and $0 \leq \left\|\frac{\tau_{21}}{\sigma_2}\right\| \leq \frac{\|\tau_{21c}\|}{R_{\perp\perp}^A}$
Definitions	$p_{\perp\|\|}^{(+)} = -\left(\frac{d\tau_{21}}{d\sigma_2}\right)_{\sigma_2=0}$ of (σ_2, τ_{21}) curve, $\sigma_2 \geq 0$	$p_{\perp\|\|}^{(-)} = -\left(\frac{d\tau_{21}}{d\sigma_2}\right)_{\sigma_2=0}$ of (σ_2, τ_{21}) curve, $\sigma_2 \leq 0$	
Parameter relationships	$R_{\perp\perp}^A = \frac{Y_C}{2\left(1 + p_{\perp\perp}^{(-)}\right)} = \frac{S_{21}}{2 p_{\perp\|\|}^{(-)}} \left(\sqrt{1 + 2 p_{\perp\|\|}^{(-)} \frac{Y_C}{S_{21}}} - 1 \right)$	$p_{\perp\perp}^{(-)} = p_{\perp\|\|}^{(-)} \frac{R_{\perp\perp}^A}{S_{21}}$	$\tau_{21c} = S_{21} \left(\sqrt{1 + 2 p_{\perp\perp}^{(-)}} \right)$

4 Numerical Results

In this section, three different problems are solved. Example 1, in Sect. 4.1, presents a validation of the proposed GA in a benchmark problem considering only one type of material. Examples 2 and 3 (Sects. 4.2 and 4.3, respectively) present optimization problems of hybrid laminates. In Example 3, the optimal design is searched under the constraint of three different failure criteria: the maximum stress (MS), Tsai-Wu (TW) and the Puck failure criterion (PFC), which are tested independently.

4.1 Example 1: Weight Minimization Under Strain, Buckling and Ply Contiguity Orientation Constraints

The purpose of this example is to validate the GA presented here by comparing it to a well-known laminated composite optimization problem [11]. Let us consider the minimal weight design of a laminated composite plate under the constraints of laminate symmetry and balance, maximum number of contiguous plies with the same orientation as well as the maximum strain and buckling. The allowable orientation angle values are 0_2, ± 45 and 90_2 degrees. Thus, the optimization problem can be stated as

$$\begin{aligned} &\text{Find}: \{\theta_k, n\}, \quad \theta_k \in \{0_2, \pm 45, 90_2\}, \quad k = 1 \,\text{to}\, n \\ &\text{Minimize}: \text{Weight} \\ &\text{Subject to}: -\text{Strain failure}(\lambda_{\text{strain}}) \\ &\qquad\qquad\quad -\text{Buckling failure}(\lambda_{\text{buckling}}) \\ &\qquad\qquad\quad -\text{Max. of 4 contiguous plies with the same orientation} \end{aligned} \tag{6}$$

where θ_k is the orientation of each stack of the laminate and n the total number of stacks. As already mentioned, each stack is composed of two layers to guarantee balance.

The plate is rectangular, simply supported and subjected to compressive in-plane loads per unit length N_x and N_y, as shown in Fig. 12. Each layer is 0.127 mm thick, and the length and width of the plate are 0.508 and 0.127 m, respectively. The classical lamination theory and the linear buckling analysis [31] are used. The buckling load factor $\lambda_{buckling}$ represents the failure buckling load divided by the applied load, and is calculated as (see, for instance, reference [33]),

$$\lambda_{\text{buckling}} = \min_{r,s} \left(\frac{\pi^2 \left[D_{11} \left(\frac{r}{a}\right)^4 + 2\left(D_{12} + 2D_{66}\right) \left(\frac{r}{a}\right)^2 \left(\frac{s}{b}\right)^2 + D_{22} \left(\frac{s}{b}\right)^4 \right]}{\left(\frac{r}{a}\right)^2 N_x + \left(\frac{s}{b}\right)^2 N_y} \right) \tag{7}$$

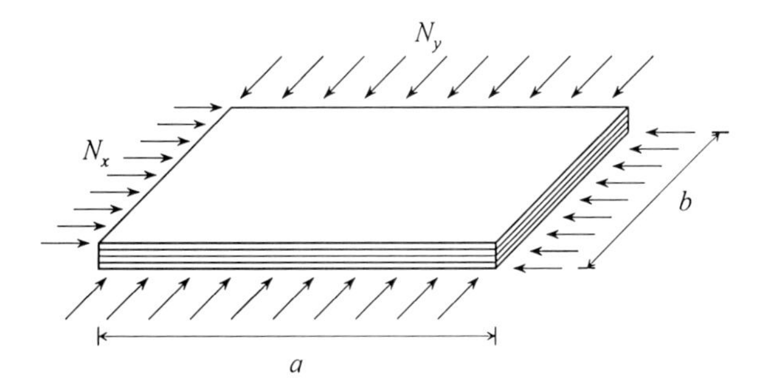

Fig. 12 Laminated composite plate subjected to in-plane compressive loads

where D_{ij} are coefficients of the laminate bending stiffness matrix, r and s determine the amount of half waves in the x and y direction, respectively, a is the plate length, b is the plate width (Fig. 12). Note that the inputs of Eq. 7 require positive values for compressive and negative values for tensile forces.

The critical strain failure factor λ_{strain} is defined as

$$\lambda_{\text{strain}} = \min_k \left[\min \left(\frac{\varepsilon_1^{ua}}{S_f \left|\varepsilon_1^k\right|}, \frac{\varepsilon_2^{ua}}{S_f \left|\varepsilon_2^k\right|}, \frac{\gamma_{12}^{ua}}{S_f \left|\gamma_{12}^k\right|} \right) \right] \tag{8}$$

where ε_i^{ua}, i $= 1, 2$ and γ_{12}^{ua}, are the allowable strains; ε_i^k, i $= 1, 2$ and γ_{12}^{ua} are the strains in the principal direction of the kth lamina and S_f is a safety factor. The critical failure factor λ_{cr} is the smallest between $\lambda_{buckling}$ and λ_{strain}.

The elastic material properties of the layers are $E_1 = 127550$ MPa, $E_2 = 13030$ MPa, $G_{12} = 6410$ MPa and Poisson's ratio $\nu_{12} = 0.3$. The ultimate strains are $\varepsilon_1^{ua} = 0.008$, $\varepsilon_2^{ua} = 0.029$ and $\gamma_{12}^{ua} = 0.015$ and the safety factor $S_f = 1.5$ is used.

The plate is analyzed under three different loading conditions: load case 1 $N_x = 2.277 \times 10^6$ N/m and $N_y = 2.846 \times 10^5$ N/m. Load case 2 has $N_x = 2.189 \times 10^6$ N/m and $N_y = 5.473 \times 10^5$ N/m. Load case 3 has $N_x = 1.716 \times 10^6$ N/m and $N_y = 8.581 \times 10^5$ N/m. For each case, 100 independent searches are performed, each of them stopped after 6000 function evaluations. The population size of the GA is 20 individuals and the probability of the genetic operators is shown in Table 3.

As the purpose of this example is to compare the results with those of Le Riche and Haftka [11], the performance evaluation of the GA follows this reference and

Table 3 Probability of the GA operators (Example 1)

Operator	Probability
Crossover	1.00
Mutation	0.10
Gene swap	1.00
Stack addition	0.05
Stack deletion	0.05

Table 4 Best designs for each load case of Example 1 (taken from reference [11])

Load case	Optimal design	Failure mode	Number of practical optima
1	$[\pm45_5/0_4/\pm45/0_4/90_2/0_2]_s$	Strain	> 13
2	$[\pm45_2/90_2/\pm45_3/0_2/\pm45/0_4/\pm45/0_2]_s$	Strain	3
3	$[90_2/\pm45_2/(90_2/\pm45)_2/\pm45_5]_s$	Buckling	13

Table 5 Price and reliability for Example 1. Comparison with reference [11]

	Load case	1	2	3
Present GA	Price	480	1120	1360
	Reliability	1.00	1.00	1.00
Reference [11][a]	Price	440	1180	1490
	Reliability	1.00	1.00	0.94

[a] Results of the so called "new GA" of reference [11]

is presented in terms of the "price of the search" and "reliability". The reliability of the algorithm is the probability it has of finding a practical optimum. A practical optimum is defined as an optimal weight design with the critical failure factor λ_{cr} within 0.1 % of λ_{cr} of the global optimum. The price of the search is the number of analyses necessary to reach 80 % reliability, i.e. to have 80% probability of finding a practical optimum. Table 4 shows the best designs found by Le Riche and Haftka [11], where all the practical optimal information can the found in detail.

The results obtained by the GA formulated in this study and their comparisons with the results of reference [11] are in Table 5. As it can be seen analysing such table, the present algorithm has achieved results that are at least as good as the one of Le Riche and Haftka [11]. The present algorithm has obtained better results regarding reliability and search price for load cases 2 and 3. The algorithm of Le Riche and Haftka [11] is still better regarding the price of the search in load case 1.

4.2 Example 2: Material Cost Minimization of a Hybrid Composite Under Buckling and Weight Constraints

In this example, the material cost minimization of a hybrid laminated composite plate is described. Two types of layers are considered: carbon-epoxy (CE) and glass-epoxy (GE). The former is lighter and stronger, while the latter has a cost advantage as the price per square meter of this laminate is about 8 times less. The laminated is subjected to symmetry and balance constraints as well as a maximum weight and a minimum buckling load factor. The allowable orientation angle values are 0_2, ±45 and 90_2 degrees. Thus, the optimization problem reads as follows

$$\begin{aligned} &\text{Find}: \{\theta_k, \text{mat}_k, n\}, \theta_k \in \{0_2, \pm 45, 90_2\}, \text{mat}_k \in \{GE, CE\}, k = 1 \text{ to } n \\ &\text{Minimize :Material cost} \\ &\text{Subject to}: - \text{Minimum buckling load factor}: \lambda_{buckling} \geq \lambda_{min} \\ &\qquad - \text{Maximum weight: 85 N} \end{aligned} \tag{9}$$

where θ_k is the orientation of each stack of the laminate and n the total number of stacks. As already mentioned, each stack is composed of two layers to guarantee balance. In this problem, each CE and GE layer is also assumed to cost 1 and 8 monetary units (m.u.), respectively.

The plate is rectangular, simply supported and subjected to compressive in-plane loads per unit length N_x and N_y, as shown in Fig. 12. Each layer is 0.127 mm thick, and the length and width of the plate are a= 0.92 m and b= 0.75 m, respectively. The classical lamination theory and the linear buckling analysis (Jones 1999) is used. The buckling load factor $\lambda_{buckling}$ is calculated as Eq. (7).

The elastic material properties of the CE layers are $E_1 = 138$ GPa, $E_2 = 9$ GPa, $G_{12} = 7.1$ GPa, Poisson's ratio $\nu_{12} = 0.30$ and mass density $\rho = 1605$ kg/m3. The GE layer elastic material properties are $E_1 = 43.4$ GPa, $E_2 = 8.9$ GPa, $G_{12} = 4.55$ GPa, Poisson's ratio $\nu_{12} = 0.27$ and mass density $\rho = 1993$ kg/m3.

The in-plane applied loads are fixed compressive values $N_x = 0.175$ N/m and $N_y = 0.175$ N/m. This problem was previously investigated by Girard [27] and the global optimum results for three different minimum buckling load factors are shown in Table 6. The underlined figures for the orientation correspond to GE stacks, and the remaining figures to CE stacks. For further comparisons, Table 6 also shows the number of function evaluations required by the algorithms proposed by Girard [27] to achieve the global optimum design. We also note that for optimum laminate configuration the CE layers are placed at the outer surface, which provides a higher bending stiffness.

In the following, we present a convergence study of this problem comparing the results of the GA developed in this study with and without the local search. The population size is equal to 20 individuals in all the analysis. The parameters used in the GA are shown in Table 7. They are the same for all the tests pursued. In the

Table 6 Optimal material cost and stacking sequence for the three different minimum buckling load factor

λ_{min}	Cost (m.u.)	$\lambda_{buckling}$	Weight (N)	Number of plies	Stacking sequence	FE[a]	FE^{b}_{local}
150	33	167.4	79.7	48	$[(\pm 45)_3\,(\underline{\pm 45})_9]_S$	14945	1426
250	55	262.4	82.6	52	$[(\pm 45)_6\,(\underline{\pm 45})_7]_S$	18345	2409
375	120	447.8	84.4	60	$[(\pm 45)_{15}]_S$	25894	1480

[a] Mean number of function evaluations to achieve the global optimum without local search, based on 50 independent runs [27]

[b] Mean number of function evaluations to achieve the global optimum using local search, based on 50 independent runs [27]

Table 7 Probability of the GA operators (Examples 2 and 3)

Operator	Probability
Crossover	1.00
Mutation	0.10
Gene swap	0.25
Stack addition	0.05
Stack deletion	0.10

material grouping local search, two of individuals among the ten best are chosen per iteration. For the neighborhood search, two individuals are investigated among the five best ones.

The study is based on 100 independent runs and the stop criterion chosen is the total number of function evaluations (FE). Table 8 shows the obtained results. The mean value and standard deviation (SD) of the buckling load factor are shown to differentiate the case that the algorithm found a solution satisfying all the constraints (defined here as a feasible solution) and when the algorithm reaches the global optimum, which is the design shown in Table 6 for each λ_{min}. Note that to calculate the mean and SD only the feasible solution values are considered.

It can be seen that in all tests the local search accelerates the convergence of the algorithm, reaching the global optimum faster.

Among the cases analyzed, $\lambda_{min} = 250$ is the hardest one for the algorithm to converge to the global optimum, once it required the highest number of function evaluations to converge. It can be also seen that, to have 100 % of probability of finding the global optimum, the local search reduced such convergence in roughly a thousand function evaluations. For the case where $\lambda_{min} = 150$, the local search achieved the global convergence in half of function evaluations.

Considering the number of function evaluations tested, $\lambda_{min} = 375$ was the easiest case to solve. In that case, the local search was not as effective or necessary as it was in the other two, meaning that the harder the optimization is, the more effective the local search may be.

Comparing the results of the three cases with those of Girard [27] (Table 6), we see that the algorithm without the local search presented here converged much faster (i.e., with a lower number of function evaluations) than the one developed there. Also, its effectiveness can be compared to the algorithm with local search of the reference. Finally, the GA developed here using the local search was the fastest among all.

4.3 Example 3: Material Cost Minimization of a Hybrid Composite Under Failure and Weight Constraints

The main purpose of this example is to pursue the material cost minimization of a hybrid laminated composite plate comparing the optimal design provided by three different failure criteria: the maximum stress (MS), Tsai-Wu (TW) and the Puck

Table 8 Convergence after 100 independent runs for different number of function evaluations as stop criterion (Example 2)

Function evaluations (FE)	300	500	1000	1500	2000	2500
$\lambda_{min} = 150$						
% of convergence[a]	0	53	96	100	100	100
λ_{cr} – mean and (SD)	–	158.2 (4.76)	165.1 (3.66)	166.8 (1.98)	167.4 (0.00)	167.4 (0.00)
% of convergence with local search	39	100	100	100	100	100
λ_{cr} – mean and (SD)	158.9 (4.83)	165.3 (3.72)	167.4 (0.00)	167.4 (0.00)	167.4 (0.00)	167.4 (0.00)
$\lambda_{min} = 250$						
% of convergence	0	11	73	91	98	100
λ_{cr} – mean and (SD)	–	255.6 (6.20)	259.1 (2.83)	261.6 (1.60)	262.3 (0.73)	262.4 (0.00)
% of convergence with local search	0	44	97	100	100	100
λ_{cr} – mean and (SD)	–	260.0 (3.68)	262.1 (0.60)	262.4 (0.00)	262.4 (0.00)	262.4 (0.00)
$\lambda_{min} = 375$						
% of convergence	61	100	100	100	100	100
λ_{cr} – mean and (SD)	412.1 (18.60)	433.5 (13.60)	447.8 (0.00)	447.8 (0.00)	447.8 (0.00)	447.8 (0.00)
% of convergence with local search	86	100	100	100	100	100
λ_{cr} – mean and (SD)	423.1 (17.80)	446.2 (2.54)	447.8 (0.00)	447.8 (0.00)	447.8 (0.00)	447.8 (0.00)

[a] Can also be interpreted as the probability of finding a feasible solution.

failure criterion (PFC). As in the preceding examples, the laminate is subjected to symmetry and balance constraints. A maximum weight constraint is also imposed in this example. Thus, the optimization problem reads as follows

$$\begin{aligned} \text{Find} &: \{\theta_k, \text{mat}_k, n\}, \theta_k \in \{0_2, \pm 45, 90_2\}, \text{mat}_k \in \{GE, CE\}, k = 1 \text{ to } n \\ \text{Minimize} &: \text{Material cost} \\ \text{Subject to} &: -\text{ Failure constraint : MS, TW or PFC} \\ & -\text{ Maximum weight : 70 N} \end{aligned} \tag{10}$$

Let us consider a carbon-epoxy square laminated plate subjected to in-plane loads per unit length $N_x = 2000$ N/mm and $N_y = -2000$ N/mm. The plate is analyzed using the classical lamination theory (see for instance, reference [31]).

Each layer is 0.1 mm thick, and the length and width of the plate are 1.0 m. The elastic material properties of the CE layers are $E_1 = 116.6$ GPa, $E_2 = 7.673$ GPa, $G_{12} = 4.173$ GPa, Poisson's ratio $n_{12} = 0.27$ and mass density $\rho = 1605$ kg/m3. The elastic material properties of the GE layers are $E_1 = 37.6$ GPa, $E_2 = 9.584$ GPa, $G12 = 4.081$ GPa, Poisson's ratio $\nu_{12} = 0.26$ and mass density $\rho = 1903$ kg/m3. The failure properties of the CE and GE laminas are shown in Table 9.

The probabilities of the GA operators are the same as in Example 2 and the optimization results are shown in Table 10. The underlined figures for the orientation correspond to GE stacks, and the remaining figures to CE stacks.

It is interesting to note that the optimum obtained followed the same pattern in every case. All layers with an orientation of 0° are made of CE, while those with an orientation of 90° are made of GE. Note that this problem is independent of the bending stiffness and, as a consequence, it is independent of the stacking sequence. Thus, if the stacking sequences shown in Table 10 are rearranged, the laminate extensional stiffness remains the same as long as the number of plies with the same orientation angle and material are kept constant. In addition, the GE laminae

Table 9 Strength properties of the layers used in Example 3

Property	Carbon-epoxy (CE)	Glass-epoxy (GE)
X_T(MPa)	2062	1134
X_C(MPa)	1701	1031
Y_T (MPa)	70	54
Y_C(MPa)	240	150
S_{12}(MPa)	105	75
E_{f1}(MPa)	230000	72000
ε_{1T}	0.0175	0.0302
ε_{1C}	0.014	0.0295
ν_{f12}	0.23	0.22
$m_{\sigma f}$	1.1	1.3
$p_{\perp\parallel}^{(+)}$	0.3	0.3
$p_{\perp\parallel}^{(-)}$	0.25	0.25

Table 10 Optimal material cost and stacking sequence of the laminate for different failure criteria (Example 3)

	Cost and weight		Failure factor (f_E)		Stacking and cost difference	
Failure Criterion	Cost (m.u.)	Weight (N)	CE	GE	Stacking sequence[a]	%[b]
PFC	144	55.57	0.81 (C)	0.95 (FF)	$\left[(0_2)_4\,(\underline{90_2})_4\right]_S$	–
PFC_fw	148	63.11	0.69 (C)	0.94 (A)	$\left[(0_2)_4\,(\underline{90_2})_5\right]_S$	2.7
TW	208	68.23	0.27	0.99	$\left[(0_2)_6\,(\underline{90_2})_4\right]_S$	30.1
MS	148	63.11	0.66	0.84	$\left[(0_2)_4\,(\underline{90_2})_5\right]_S$	2.7

[a] Any order of this combination of orientation and material gives the same response.
[b] Relative weight difference (percentage difference in relation to the weight obtained using the PFC) predicted by the failure criteria.

were the closest to failure. The cheapest structure was obtained using the PFC, while the TW criterion resulted in a material cost over 30% higher and yielded the heaviest structure.

Table 10 also shows the maximum failure factor for the CE and GE laminae. The TW criterion yielded the largest gap between the maximum failure efforts for the two different materials at the optimum. Again, the PFC provides not only the failure effort, but also the expected failure mode of the structure. For example, the PFC predicts that the most probable failure mode is FF, while PFC_fw predicts IFF (mode A). From the results obtained, we note that each failure criterion yielded a different optimum. This reinforces the idea that the failure criterion significantly modifies the optimal design. Thus, when optimizing laminated composite structures, the choice of a failure criterion corresponding to the real behavior of the structure is crucial for both economy and safety.

5 Conclusions

In this chapter, a genetic algorithm was developed to pursue the optimization of hybrid laminated composite structures. The GA was chosen as an optimization tool because of its ability to deal with non-convex, multimodal and discrete optimization problems, of which the design of laminated composites is an example. First, the developed algorithm was validated by comparing its results to those obtained from the literature. Then, two more examples were solved. In one of them, aiming to analyze the influence of the failure criterion over the optimal design, maximum stress, Tsai-Wu and Puck failure criteria were used, independently, as constraint in the optimization problem and the results yielded by them were compared.

The results of this study show that the developed algorithm converges faster than the one found in the literature and that the local search accelerates the convergence. Moreover, the harder the problem is, the more effective the local search is.

Regarding the different failure criteria, it was found that each criterion yields a different optimal design. Therefore, when optimizing laminated composite structures, the choice of a failure criterion corresponding to the real behavior of the structure is crucial for both economy and safety.

References

1. Kim, H.A., Kennedy, D., Gürdal, Z.: Special issue on optimization of aerospace structures. Struct Multidiscip Optim **36**, 1–2 (2008)
2. Nagendra, S., Haftka, R.T.Gürdal, Z, : Stacking sequence optimization of simply supported laminates with stability and strain constraints. AIAA J **30**, 2132–2137 (1992)
3. Le Riche, R., Haftka, R.: Optimization of laminate stacking sequence for buckling load maximization by genetic algorithm. AIAA J **31**, 951–956 (1993)
4. Nagendra, S., Jestin, D., Gürdal, Z., Haftka, R., Watson, L.: Improved genetic algorithm for the design of stiffened composite panels. Comput Struct **58**, 543–555 (1994)
5. Todoroki, A., Haftka, R.: Stacking sequence optimization by a genetic algorithm with a new recessive gene like repair strategy. Composites Part B **29**, 277–285 (1998)
6. Liu, B., Haftka, R., Akgun, M., Todoroki, A.: Permutation genetic algorithm for stacking sequence design of composite laminates. Comput Meth Appl Mech Eng **186**, 357–372 (2000)
7. Lopez, R.H., Luersen, M.A., Cursi E.S.: Optimization of laminated composites considering different failure criteria. Composites: Part B **40**, 731–740 (2009)
8. Callahan, K.J., Weeks, G.E.: Optimum design of composite laminates using genetic algorithm. Compos Eng **2**, 149–160 (1992)
9. Groenwold, A., Haftka, R.: Optimization with non-homogeneous failure criteria like Tsai-Wu for composite laminates. Struct Multidiscip Optim **32**, 183–190 (2006)
10. Seresta, O., Gürdal, Z., Adams, D.B., Watson, L.T.: Optimal design of composite wing structures with blended laminates. Composites Part B **38**, 469–480 (2007)
11. Le Riche, R., Haftka, R.: Improved genetic algorithm for minimum thickness composite laminate design. Compos Eng **5**, 143–161 (1995)
12. Naik, G.N., Gopalakrishnan, S., Ganguli, R.: Design optimization of composites using genetic algorithms and failure mechanism based failure criterion. Compos Struct **83**, 354–367 (2008)
13. Diaconu, C.G., Sato, M., Sekine, H.: Layup optimization of symmetrically laminated thick plates for fundamental frequencies using lamination parameters. Struct Multidiscip Optim **24**, 302–311 (2002)
14. Karakaya, S., Soykasap, O.: Natural frequency and buckling optimization of laminated hybrid composite plates using genetic algorithm and simulated annealing. Struct Multidiscip Optim **43**, 61–72 (2011)
15. Sadr, M.H., Bargh, H.G.: Optimization of laminated composite plates for maximum fundamental frequency using Elitist-Genetic algorithm and finite strip method. J Global Optim **54**, 707–728 (2012)
16. Rahul, Sandeep G., Chakraborty, D., Dutta, A.: Multi-objective optimization of hybrid laminates subjected to transverse impact. Compos Struct **73**, 360–369 (2006)
17. Murugan, M.S., Suresh, S., Ganguli, R., Mani, V.: Target vector optimization of composite box beam using real-coded genetic algorithm: a decomposition approach. Struct Multidiscip Optim **33**, 131–146 (2007)

18. Corvino, M., Iuspa, L., Riccio, A., Scaramuzzino, F.: Weight and cost oriented multi-objective optimisation of impact damage resistant stiffened composite panels. Comput Struct **87**, 1033–1042 (2009)
19. Abouhamze, M., Shakeri, M.: Multi-objective stacking sequence optimization of laminated cylindrical panels using a genetic algorithm and neural networks. Compos Struct **81**, 253–263 (2007)
20. Badalló, P., Trias, D., Marín, L., Mayugo, J.A.: A comparative study of genetic algorithms for the multi-objective optimization of composite stringers under compression loads. Composites Part B **47**, 130–136 (2013)
21. Honda, S., Igarashi, T., Narita, Y.: Multi-objective optimization of curvilinear fiber shapes for laminated composite plates by using NSGA-II. Composites Part B **45**, 1071–1078 (2013)
22. Antonio, C.: A multilevel genetic algorithm for optimization of geometrically nonlinear stiffened composite structures. Struct Multidiscip Optim **24**, 372–386 (2001)
23. Antonio, C.: A hierarchical genetic algorithm with age structure for multimodal optimal design of hybrid composites. Struct Multidiscip Optim **31**, 280–294 (2006)
24. Puck, A., Schürmann, H.: Failure analysis of FRP laminates by means of physically based phenomenological models. Compos Sci Technol **58**, 1045–1067 (1998)
25. Puck, A.: Festigkeitsanalyse von Faser-Matrix-Laminaten: Modelle für die Praxis (Strength Analysis of Fiber-Matrix/Laminates: Models for Design Practice). Carl-Hanser- Verlag, Munich (1996)
26. Lopez, R.H., Luersen, M.A., Cursi, J.E.S.: Optimization of Hybrid Laminated Composites using a Genetic Algorithm. J Braz Soc Mech Sci & Eng **31**, 269–278 (2009)
27. Girard, F.: Optimisation de Stratifiés en Utilisant un Algorithme Génétique (Optimization of Laminates using a Genetic Algorithm). Sciences and Engineering Faculty, University of Laval, Quebec, Canada, Master's Dissertation (2006)
28. Goldberg, D.E.: Genetic algorithms in search, optimization and machine learning. Addison-Wesley, Boston (1989)
29. Schmitt, L.M.: Theory of genetic algorithms. Theor Comput Sci **259**, 1–61 (2001)
30. Puzzi, S., Carpinteri, A.: A double-multiplicative dynamic penalty approach for constrained evolutionary. Struct Multidiscip Optim **35**, 431–445 (2008)
31. Jones, R.M.: Mechanics of Composite Materials. Taylor and Francis, Philadelphia (1999)
32. Daniel, I.M.: Failure of Composite Materials. Strain **43**, 4–12 (2007)
33. Gürdal, Z., Haftka, R.T., Hajela, P.: Design and Optimization of Laminated Composite Materials. John Wiley & Sons, New York (1999)

Delamination Diagnosis in Composite Beam Using AIS and BGA Algorithms Based on Vibration Characteristics

B. Mohebbi, F. Abbasidoust, M. M. Ettefagh and H. Biglari

Abstract In this study vibration-based delamination detection is achieved using the artificial immune system (AIS) method. The approach is based upon mimicking immune recognition mechanisms that possess features such as adaptation, evolution, and immune learning. The identification of the delamination location and its length in the composite beam is formulated as an optimization problem. The cost function is based on differences between analytical or experimental natural frequencies and predicted natural frequencies by AIS method. Analytical natural frequencies of delaminated beam are obtained from Euler-Bernoulli beam theory with constrained delamination mode. Also, in this paper the binary genetic algorithm (BGA) was applied to compare the predicted locations and lengths with those obtained from AIS method. Errors of predicted location and length are 2.16 % and 0.1968 % respectively, using the AIS and these values become −3.83 % and 1.76 % for the BGA. The proposed approach (AIS) showed significantly better performance in detecting failures in comparison with the other method (BGA). In addition, detection accuracy and prediction errors, calculated with variance account for (VAF) and mean square error (MSE) concepts are compared with different artificial neural networks (ANNs) methods. To investigate the accuracy of the proposed method, some experimental results were obtained. A laser vibrometer was used to identify natural frequencies change in delaminated carbon-fiber-reinforced polymer (CFRP) composite beam case study. The average error values of predicted location and length in experiment test are 18.7 % and 9.8 %, respectively.

B. Mohebbi · F. Abbasidoust · M. M. Ettefagh (✉) · H. Biglari
Mechanical Engineering Department, University of Tabriz, 51666-16471Tabriz, Iran
e-mail: ettefagh@tabrizu.ac.ir

P. A. Muñoz-Rojas (ed.), *Optimization of Structures and Components*,
Advanced Structured Materials 43, DOI: 10.1007/978-3-319-00717-5_5,

1 Introduction

Damage detection in composite structures is an important issue because of their increasingly use in the construction of aerospace, civil, marine, automotive and other high performance structures. Delamination is one of the major damage modes in laminated composites due to their weak interlaminar strength. Delaminations may arise during fabrication or service-induced strains, such as incomplete wetting, air entrapment, impact of foreign objects, exposure to unusual level of excitation, etc [1, 2]. Delaminations are embedded within the composite structures so they may not be visible or barely visible on the surface but they can reduce stiffness and strength of the structures [3, 4]. A reduction in stiffness reduces natural frequency and may cause resonance if the reduced frequency is close to the working frequency. Therefore it is important to understand the influence of the delaminations on the vibration characteristics of the structures [5]. Damage detection in structures is one of the common topics that have received growing interest in research communities [6].

While a number of damage detection and localization methods have been proposed, in this paper, a novel damage diagnosing method based on AIS algorithm has been developed, which incorporates several major characteristics of the natural immune system [7]. The identification of the delamination location and length in the cantilever beam, as a case study, is formulated as an optimization problem [8]. For this purpose, the damage patterns are represented by feature vectors that are extracted from the structure's dynamic response measurements. Also, the possible changes in the natural frequencies of the structure are utilized as a feature vector because frequencies can be measured more easily than other vibration parameters such as mode shapes as well as being less critically affected by experimental errors.

The selective and adaptive features of the proposed algorithm allow the AIS to evolve its antibodies towards the goal of minimizing the cost function. The performance of the presented structure damage detector has been validated using a model of damaged (delaminated) beam. For modeling the delaminated-beam composite structure an analytical model of a delaminated cantilever beam is utilized and natural frequencies are obtained through numerical methods [8]. Also, the frequencies have been obtained by experimental tests for a CFRP composite beam, which its model has been updated by BGA. Both these frequencies obtained by experiment and theory have been used as the AIS algorithm's input. The results show that experiments validate the integrity of our method.

2 Natural and Artificial Immune Systems

The biological immune system is a complex adaptive system that has evolved in vertebrates to protect them from invading pathogens. The biological immune system can be envisioned as a multilayer protection system, where each layer provides different types of defense mechanisms for detection, recognition and responses. Thus, three main layers include the anatomic barrier, innate immunity (nonspecific) and adaptive immunity (specific). Innate immunity and adaptive immunity are interlinked and influence each other. Once adaptive immunity recognizes the presence of an invader, it triggers two types of responses: humoral immunity and cell-mediated (cellular) immunity, which act in a sequential fashion. Innate immunity is directed against any pathogen. If an invading pathogen escapes the innate defenses, then the body can launch an adaptive or specific response against a particular type of foreign agent [7].

During the last decade, based on principles of the immune system, a new paradigm, called AIS, has been employed for developing interesting algorithms in many fields such as pattern recognition, computer defense, fault detection, optimization, and others. The field of AIS is progressing slowly and steadily as a branch of CI.

In recent years, some different AIS-based algorithms have been developed. Among various mechanisms in the biological immune system that are explored as AIS's, negative selection, immune network model and Clonal selection are still the most discussed models [8]. Following sections defines the main selection methods used in the AIS algorithm.

2.1 Clonal Selection

The Clonal Selection Principle describes the basic features of an immune response to an antigenic stimulus. It establishes the idea that only those cells that recognize the antigen proliferate, thus being selected against those that do not [9]. The main features of the Clonal Selection Theory are that:

- The new cells are copies of their parents (clone) subjected to a mutation mechanism with high rates (somatic hypermutation).
- Elimination of newly differentiated lymphocytes carrying self-reactive receptors.
- Proliferation and differentiation on contact of mature cells with antigens.

The clonal selection and affinity maturation based algorithms such as CLONALG algorithm have several interesting features:

(1) Population size dynamically adjustable.
(2) Exploitation and exploration of the search space.
(3) Location of multiple optima.
(4) Capability of maintaining local optima solutions.
(5) Defined stopping criterion.

2.2 Negative Selection

One of the purposes of the immune system is to recognize all cells (or molecules) within the body and categorize those cells as self or non-self. The non-self cells are further categorized in order to induce an appropriate type of defensive mechanism.

The immune system learns through evolution to distinguish between foreign antigens (bacteria, viruses, etc.) and the body's own cells or molecules. The purpose of negative selection is to provide tolerance for self cells. It deals with the immune system's ability to detect unknown antigens while not reacting to the self cells [10].

Two important aspects of a negative selection algorithm are:

(1) The target concept of the algorithm is the complement of a self set.

(2) The goal is to discriminate between self and non-self patterns, but only samples from one class are available (one class learning) [7].

3 AIS-Based Structure Damage Detection

The fault detection system is designed using concepts derived from the natural immune system. The component correspondence between the natural immune system and the AIS-based structure damage detection method is cited in the next subsection. The AIS algorithm in our implementation is used to damage detection by defining a cost function $C(.,.)$,

$$C(X, L) = \sum_{i=1}^{4} abs(f_i - f_i^*) \tag{1}$$

Which should be minimized based upon the calculated and corresponding to unknown natural frequencies. In Eq. (1) X is the location of the delamination from one end, and L is the length of the delamination. f_i's are the first three natural frequencies, which are functions of X and L, and are calculated from the delaminated-beam model, f_i^*'s are the first three natural frequencies which are applied to our delamination detection system as inputs. A zero cost value indicates an exact match between the corresponding frequencies. In this section, the modified immune optimization algorithm is discussed.

3.1 Immune Terminologies and Representations

This section defines the major components and parameters used in the AIS algorithm. Antigen: each peak of the function to be optimized.

Antibody: an individual of the population represented as a real-valued attribute string in the Euclidean shape-space.

Fitness: affinity between the antibody and Ag, an index to measure the goodness of the antibody.

Affinity: Euclidean distance between two antibodies.

Clone: offspring antibodies that are identical to the parent antibody.

Mutated clone: a clone that has undergone somatic mutation.

Memory cell: the antibodies with higher fitness and antigenic affinities are selected to become memory cells with long life spans.

Negative selection: if there is any antibodies in memory cells, whose affinity with Abi is higher than a defined limit this Ab_i is discarded.

3.2 The Algorithm

The algorithm that used in purposed AIS method follows the steps shown in Fig. 1 These steps described more precisely below:
Step1: Generate n antibodies randomly within the range (0, 1).

$$\begin{aligned} p &= \{p_1, p_2, p_3, \ldots, p_n\} \\ p_1 &= (x_{11}, x_{12}, x_{13}, \ldots, x_{1d}) \\ p_2 &= (x_{21}, x_{22}, x_{23}, \ldots, x_{2d}) \\ &\vdots \\ p_n &= (x_{n1}, x_{n2}, x_{n3}, \ldots, x_{nd}) \end{aligned}$$

Step2: Generate nc of clones for each antibody Ab_i in the population.

$$\begin{aligned} N_C &= n \times n_c \\ C &= \{C_1, C_2, C_3, \ldots, C_n\} = \{q_1, q_2, q_3, q_4, q_5, \ldots, q_{N_c}\} \\ C_1 &= \overbrace{p_1, p_1, p_1, \ldots, p_1}^{n_c} \\ C_2 &= \{p_2, p_2, p_2, \ldots, p_2\} \\ &\vdots \\ C_n &= \{p_n, p_n, p_n, \ldots, p_n\} \end{aligned}$$

Step3: Mutate each clone. The mutation rate has inverse relationship with the number of iteration of parent invariance.

$$\begin{cases} c_i^* = q_i + \mu \times \text{rand} & \text{for} \quad i = 2, 3, 4, \ldots, nc, nc+2, nc+3, \ldots, 2nc, 2nc+2, \ldots \\ c_i^* = q_i & \text{for} \quad i = 1, nc+1, 2nc+1, 3nc+1, \ldots \end{cases}$$
$$C^* = \{c_1^*, c_2^*, c_3^*, \ldots, c_{N_c}^*\}$$

Rand is a number between (0, 1).

Step4: Decode the antibodies within the domain of variables of function and determine the fitness of all the individuals in the population.

Step5: For each clone, select only the one with the highest fitness to generate a new population.

Step6: recognizing the answers and enter them in memory cell. If a member of population does not change in N_{mem} iterations, it will be saved in memory cell.

Step7: using negative selection to replace random member instead of self antibodies.

Step8: Repeat Step 2–5 until stopping criteria's is satisfied.

- Maximum number of iterations allowed.
- Maximum number of iterations that algorithm can continue without finding a new answer.
- Maximum number of answers that algorithm can find.

4 The Genetic Algorithm Method

GAs are a problem-solving heuristic technique inspired by biological evolution in the same way that AIS inspired by the human immune system. They have been successfully applied to a wide range of real-world problems of significant complexity. Each GA operates on a population of artificial chromosomes. These are strings in a finite alphabet (usually binary). Each chromosome represents a solution to a problem and has a fitness, a real number which is a measure of how good a solution it is to the particular problem. In the present paper we also use BGA in order to compare the prediction results with results obtained using AIS algorithm. Cost function in BGA algorithm is the same as it was in AIS method and defined with Eq. (1). In our approximation, the binary genetic algorithm is used with two point crossover method and mutation. After several examinations the best answer was received when the mutation number was 0.33. In each run the mutation rate and the generation number were 0.024 and 250, respectively.

5 Delaminated Composite Beam Modeling

In order to detect delamination locations/size, it's necessary to obtain the analytical solution to the vibration of composite beams with different Delaminations. The delaminated beam considered as $3n + 1$ Euler-Bernoulli classical beam (with $Li \gg hi$) [11, 12] which are connected at the delamination boundaries, where n is the number of delaminations (geometry of beam in shown in Fig. 2). In the present paper, the 'constrained mode' is considered. Using Euler-Bernoulli beam theory, the governing equations for intact beam section are [13]:

$$D_i \frac{\partial^4 w_i}{\partial x^4} + \rho_i A_i \frac{\partial^2 w_i}{\partial t^2} = 0, \tag{2}$$

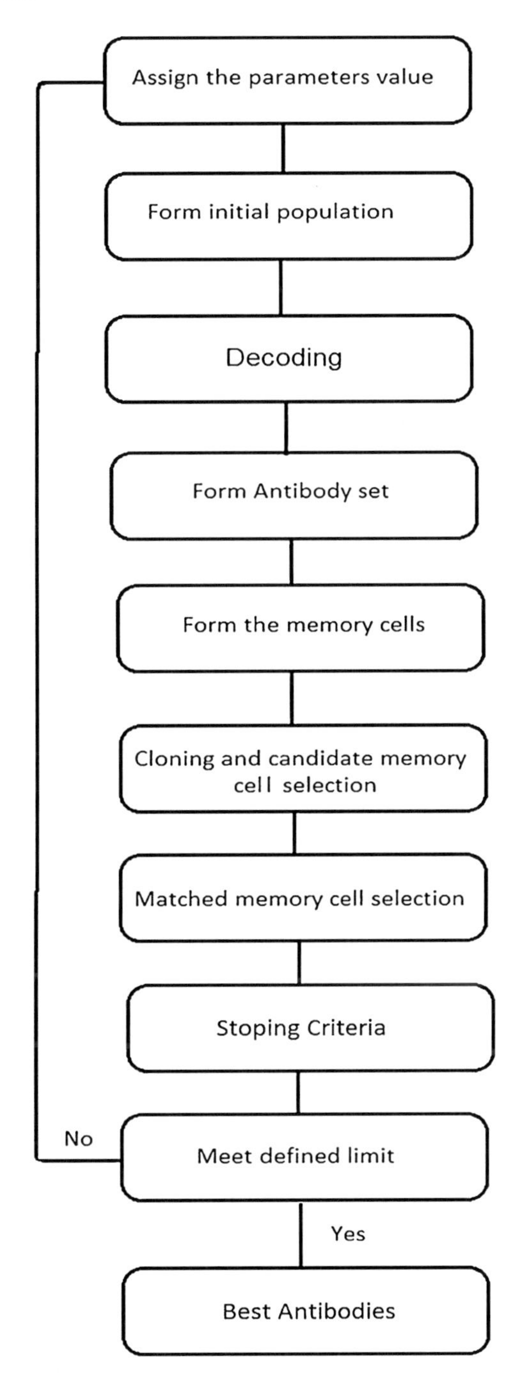

Fig. 1 Flowchart of AIS algorithm

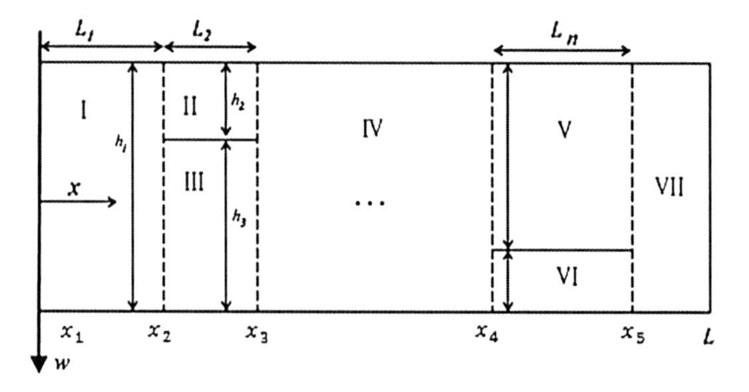

Fig. 2 Delaminated beam geometry

where D_i is the reduced bending stiffness of the i—the beam, ρ_i is the density of material, A_i is the cross sectional area; x is the axial coordinate and t is the time. The reduced bending stiffness is as follows [14]:

$$D_i = D_{11}^i - \frac{(B_{11}^i)^2}{(A_{11}^i)} \tag{3}$$

where

$$D_{11}^{(i)} = \frac{b}{3}\sum_{k=1}^{n_i}\left(\overline{Q_{11}^k}\right)_k\left(z_k^3 - z_{k-1}^3\right), \tag{4}$$

$$B_{11}^{(i)} = \frac{b}{2}\sum_{k=1}^{n_i}\left(\overline{Q_{11}^k}\right)_k\left(z_k^2 - z_{k-1}^2\right), \tag{5}$$

$$A_{11}^{(i)} = b\sum_{k=1}^{n_i}\left(\overline{Q_{11}^k}\right)_k(z_k - z_{k-1}), \tag{6}$$

$$\overline{Q_{11}^k} = Q_{11}^k\cos^4\phi + 2\left(Q_{12}^k + 2Q_{44}^k\right)\sin^2\phi\cos^2\phi + Q_{22}^k\sin^4\phi, \tag{7}$$

$$Q_{11} = \frac{E_{11}}{1-\nu_{12}\nu_{21}}\, Q_{22} = \frac{E_{22}}{1-\nu_{21}\nu_{12}}\, Q_{66} = G_{12}\,\nu_{12} = \frac{\nu_{12}E_{22}}{E_{11}} \tag{8}$$

where $A_{11}^{(i)}$ is the extensional stiffness, $D_{11}^{(i)}$ is the bending stiffness, $B_{11}^{(i)}$ is the coupling stiffness, $\overline{Q_{11}^k}$ is the coefficient stiffness of the lamina, b is the width, W_{11} and E_{22} are the longitudinal and transverse young's moduli, respectively, ν_{12} and ν_{21} are the longitudinal and transverse Poisson's ratio, G_{12} is the in-plane shear modulus, respectively, ϕ is the angle of kth lamina orientation where $2k$ and $2k-1$ are the locations of the kth lamina with respect to the midplane of ith beam.
The Eq. (2) is valid for intact beam. According to the concept of constrained model the delaminated Sections are forced to vibrate together, therefore governing equation for beams 1–4 (see Fig. 3) are:

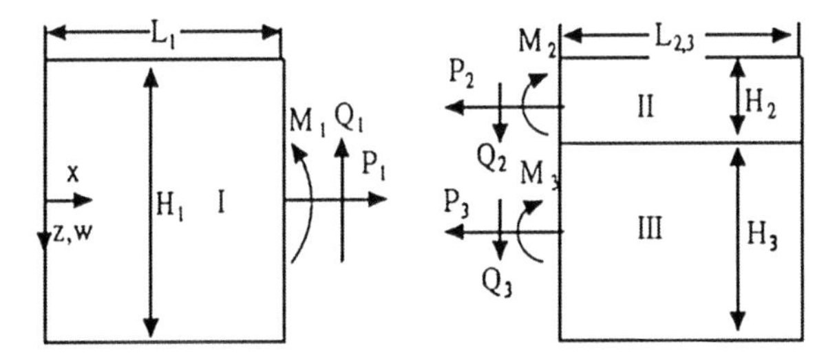

Fig. 3 Bending moment and shear force at the delamination boundary

$$D_1 \frac{\partial^4 w_I}{\partial x^4} + \rho_1 A_1 \frac{\partial^2 w_I}{\partial t^2} = 0 \tag{9}$$

$$(D_3 + D_2) \frac{\partial^4 w_{II}}{\partial x^4} + (\rho_2 A_2 + \rho_3 A_3) \frac{\partial^2 w_{II}}{\partial t^2} = 0 \tag{10}$$

and

$$D_4 \frac{\partial^4 w_{IV}}{\partial x^4} + \rho_4 A_4 \frac{\partial^2 w_{IV}}{\partial t^2} = 0. \tag{11}$$

For free vibrations, the response can be assumed as follows:

$$w_i(x, t) = W_i(x) \sin(\omega t) \tag{12}$$

Where ω denotes natural frequency and $W_i(x)$ is the mode shape of ith beam section substituting Eq. (12) into Eq. (2) and eliminating the trivial solution $\sin(\omega t) = 0$ one can obtain the general solution of the differential Eq. (2) as:

$$W_i(x) = C_i \cos(\lambda_i \frac{x}{L}) + S_i \sin(\lambda_i \frac{x}{L}) + CH_i \cosh(\lambda_i \frac{x}{L}) + SH_i \sinh(\lambda_i \frac{x}{L}), \tag{13}$$

where

$$\lambda_i^4 = \frac{\omega^2 \rho_i A_i}{D_i} L^4 \tag{14}$$

Unknown coefficients C_i, S_i, SH_i, CH_i can be determined from boundary and continuity conditions. The continuity conditions for the deflection, slope, shear (Q) and bending moments (M) at $x = x_2$ are as follows (see Fig. 3) [11]:

$$W_1 = W_{II} \tag{15}$$

$$W_1' = W_{II}' \tag{16}$$

$$D_1 W_I''' = (D_2 + D_3) W_{II}''' \tag{17}$$

$$M_1 = M_2 + M_3 - \frac{1}{2} P_2 (H_1 - H_2) + \frac{1}{2} P_3 (H_1 - H_3) \tag{18}$$

where

$$Q_i = -D_i W_i''' \quad M_i = -D_i W''' \tag{19}$$

The axial forces P_i can be established from compatibility between stretching/ shortening of the delaminated layers and axial equilibrium which results

$$\frac{P_3 L_2}{A_{11}^3} - \frac{P_2 L_2}{A_{11}^2} = (W_1'(x_2) - W_4'(x_3)) \frac{H_1}{2} \tag{20}$$

$$P_1 = P_2 + P_3 = 0 \tag{21}$$

We can obtain continuity equations for bending moment with substituting Eq. (20) and Eq. (21) into Eq. (18)

$$D_1 W_1''(x_2) + \frac{H_1^2}{L_2} \frac{A_{11}^{(2)} A_{11}^{(3)}}{A_{11}^{(2)} + A_{11}^{(3)}} (W_1'(x_2) - W_4'(x_3)) - (D_2 + D_3) D_1 W_2''(x_2) \tag{22}$$

The boundary and continuity conditions provide a coefficient matrix with $8n + 4$ orders where n is the number of delaminations. A non trivial solution for the coefficients exists only when determinant of the coefficient matrix vanishes.

6 Numerical Results

The method proposed in this work is used to analyse a cantilever beam with a single delamination, which was studied by Shen and Grady [15]. The beam is made of $T300/934$ graphite/epoxy cantilever beam with a $[0^\circ/90^\circ]_{2s}$ stacking sequence. The dimensions of the 8-ply beam are $127 \times 12.7 \times 1.016\,\text{mm}^3$. The material properties for the lamina are: $E_{11} = 134GPa$, $E_{22} = 10.3GPa$, $G_{12} = 5GPa$, $_{12} = 0.33$ and $\rho = 1.48 \times 103\,\text{kg/m}^3$.
All delaminations are at the midspan and the lengths are 25.4mm, 50.8mm, 76.2mm and 101.6 mm. The first frequencies calculated for the composite beam with cantilever boundary conditions, $(h_2/h_1 = 0.5)$ are shown in Table 1. This Table illustrates that natural frequencies decreased as Delamination length increased and there is a good agreement between the frequencies calculated by the present model and experimental [15], analytical [12] and FEM results [15].

Table 1 Primary frequencies for cantilever composite beam

Delamination lengh (mm)	Present (Hz)	Shu and Della [12]	Shen and Grady [15]	Lou and Hangud [16]
0.0	82.01	81.88	82.04	81.86
25.4	80.02	80.47	80.13	81.84
50.8	74.45	75.36	75.29	76.81
76.2	65.14	66.14	66.94	67.64
101.6	54.81	55.67	57.24	56.95

Table 2 The input natural frequencies

Test point no.	Desired value (mm)		Input natural frequencies(Hz)		
	X	L	1st	2nd	3rd
1	50.8	25.4	80.0288	481.6375	879.6543
2	38.1	50.8	74.4523	447.4076	851.9948
3	25.4	76.2	65.1414	385.3599	752.4364
4	12.7	101.6	54.8101	308.8741	713.5140

Table 3 Predicted values by AIS

Perdicition (mm)		Error (mm)		Error (%)	
X	L	X	L	X	L
Using Artificial Immune system					
50.3	25.4	−0.5	0	−0.98	0
39.3	50.7	1.2	−0.1	3.14	−0.1968
25.4	76.2	0	0	0	0
12.7	101.6	0	0	0	0
Average		0.7	−0.1	2.16	0.1968

The input natural frequencies are shown in Table 2. Because there are several Locations and Length with the same first natural frequency so using the first three natural frequencies reduces errors.

Table 3 presents predicted test points of Delamination in AIS algorithm for different locations and lengths. The prediction error is calculated from the following formula:

$$\%\text{Error} = \frac{\text{predicted value} - \text{actual value}}{\text{actual value}} \times 100(\%) \tag{23}$$

For each test point the algorithm runs from five different initial random points and the answers with the lowest cost value is selected as the best answer. Table 3 illustrates that the algorithm is more accurate in predicting lengths (L) than locations (X). Also it shows that as the length of delamination increases, the error in predicting both location and length decreases.

Table 4 MSE values for AIS method

Desired value (mm)		Mean-square error (MSE)	
X	L	Position (X)	Lenghth (L)
50.8	25.4	9.69	0.05
38.1	50.8	3.38	0.02
25.4	76.2	0.09	0.1
12.7	101.6	0	0.02

Table 5 VAF values for AIS method

Test point no.	VAF values for position	VAF values for length
1	98.7417	99.9947
2	99.8661	99.9994
3	99.9678	99.9998
4	99.9380	99.9981
5	98.7520	99.9947
Average	99.4531	99.4511

Mean-square error values that calculated for AIS method are shown in Table 4. The mean-square network error was calculated as follows:

$$MSE = \sum_{i=1}^{n} (y_a - y_p)^2 \tag{24}$$

Where n shows number of algorithm run, y_a, y_p denote actual value and predicted value of the target. The VAF is shown in Table 5. The variance accounts for (VAF) have been calculated for each test point [17].

$$VAF = \left[1 - \frac{\text{var}(y_a - y_p)}{\text{var}(y_a)}\right] \cdot 100 \tag{25}$$

where var denotes variance, VAF shows the overall performance and ideally is equal to 100 %. VAF is calculated for several ANN methods by H. Hein and L. Feklistova they had used The back-propagation neural network (BPNN) [18]. The comparison and validation of the VAF for AIS-based algorithm results, is done using these results in Table 6.

Also length and location of delaminations extracted using binary genetic algorithm (BGA) and results for our four test points are shown in Table 7. We see that average and maximum errors for BGA are more than corresponding errors for AIS algorithm. Figure 4a–d. shows cost values versus memory cell number for our four study cases. We have sorted cost values and these graphs indicate that which memory cell has the minimum cost value. For example in case (a) minimum cost value occurs in memory cell number #11.

Figure 5 shows a three dimensional plot of cost value versus parameters (Position and Length) for test point 1 (see Table 2). It shows that there is a local minima near optimization goal ($X = 50.8$(mm), $L = 25.4$(mm)) where cost value is equal to

Table 6 VAF values for delamination length using AIS (present) and ANN methods [18]

Method	VAF values for length
ANN-Resilient	
Fr	0.989
H8	0.595
ANN-Fletcher-Reeves	
Fr	0.968
H8	0.646
ANN-Palak-Ribiere	
Fr	0.991
H8	0.496
ANN-Powell-Beale	
Fr	0.987
H8	0.631
ANN-Levenberg-Marquadt	
Fr	0.983
H8	0.435
ANN-Bayesian	
Fr	0.986
H8	0.888
AIS (present)	0.994

Table 7 Predicted values by BGA

Perdicition (mm)		Error (mm)		Error (%)	
X	L	X	L	X	L
Using binary genetic algorithm					
43.9	25.4	−6.9	0	−13.58	0
39.7	49.3	1.6	−1.5	4.19	−2.95
22	79.8	−3.4	3.6	−13.38	4.72
15.9	101.6	3.2	0	18.93	0
Average		−5.5	2.1	−3.83	1.76

zero, also algorithm find several answers in this area so there is a good chance that cost value becomes zero.

7 Experimental Setup and Test

We carried out two experiments to investigate the accuracy of the present model and measurements errors. In the first experiment the test beam should be updated because material properties that used in analytical solution could differ from ones applied in experimental specimen. For this purpose, the first three natural frequencies of an intact beam are measured and results are used to update the material properties of

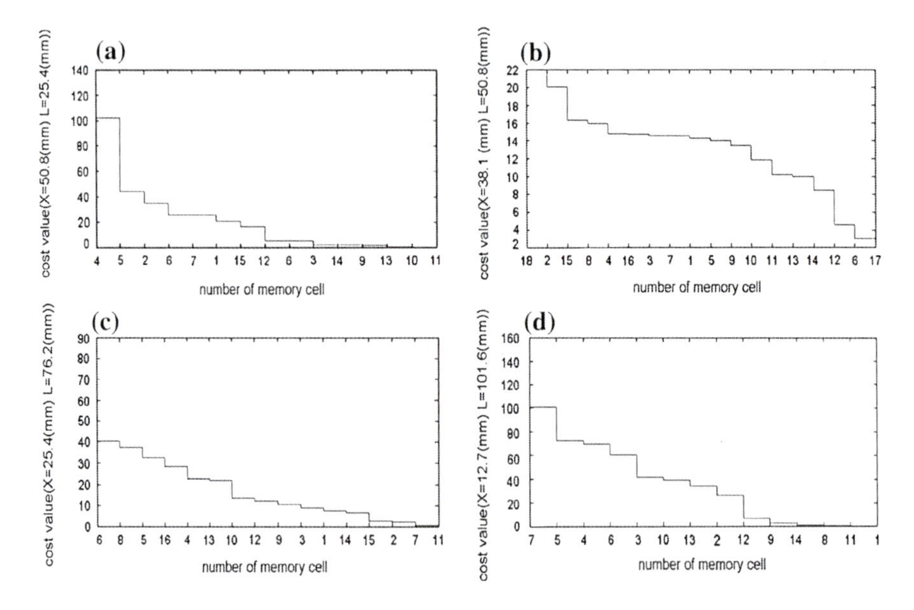

Fig. 4 (**a**–**d**) Cost values for four different case studies versus number of memory cell

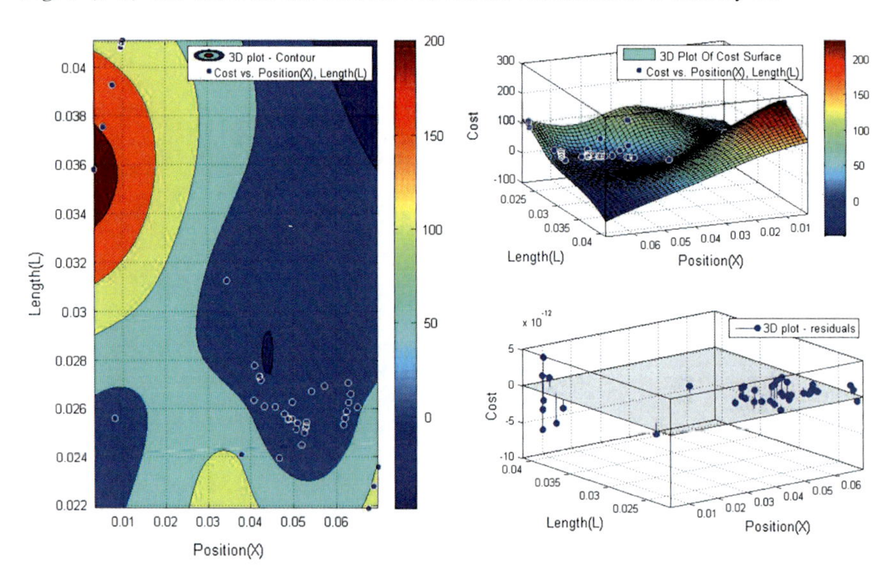

Fig. 5 Three dimensional plots (contour, surface, residuals) of cost value versus position (X) And length (L)

Table 8 Measured natural frequencies for experimental cases

Test case	Location (mm)	Length (mm)	Natural frequencies (Hz)		
			1st	2nd	3rd
Delaminated	50.8	25.4	64.8	349	1050
Intact		68	417	1221	

Table 9 Updated properties using AIS algorithm

E_{11} (GPa)	E_{22} (GPa)	G_{12} (GPa)	v_{12}	ρ(kg/m3)	
Initial properties	134	10.3	5	0.33	1.48 × 103
Updated properties	110.55	11.86	4.6	0.28	1.8 × 103

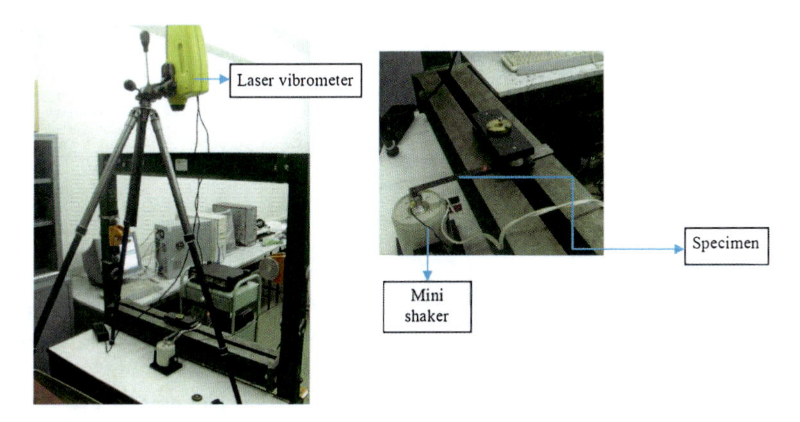

Fig. 6 Experimental setup

our model with AIS method. Optimization targets were: E_{11}, E_{22}, G_{12}, $_{12}$ and ρ. The results of natural frequency measurements for intact experimental cases are shown in Table 8 and updated material properties are shown in Table 9. In addition the experimental setup is shown in Fig. 6.

In the second experiment, a Teflon strip cured into the center of the laminate for damage simulation in the center of the beam. We measured three lowest natural frequencies of the laminated beam with delamination length of 25.4 mm and in the midspan of the beam with $(h_2/h_1 = 0.5)$ and we use these frequencies to predict delamination location (X) and length (L). Finally we calculated errors in the same way as before. The estimation results are shown in Table 10 in which the simulation test results, obtained from the model as it is illustrated in Table 3, are better than the experimental ones. The errors are because of some modeling errors, experimental uncertainties, and measurement errors.

It should be noted that for natural frequency estimation in above mentioned experiments, the beam was excited by a mini shaker (JZK Sinocera) at the distance of 127 mm from the fixed end. The dynamic response of the beam was measured using

Table 10 Actual and estimation delamination length and location for experimental case

Exact value(mm)		Perdicition (mm)		Error (mm)		Error (%)	
X	L	X	L	X	L	X	L
50.8	25.4	60.3	27.9	9.5	2.5	18.7	9.8

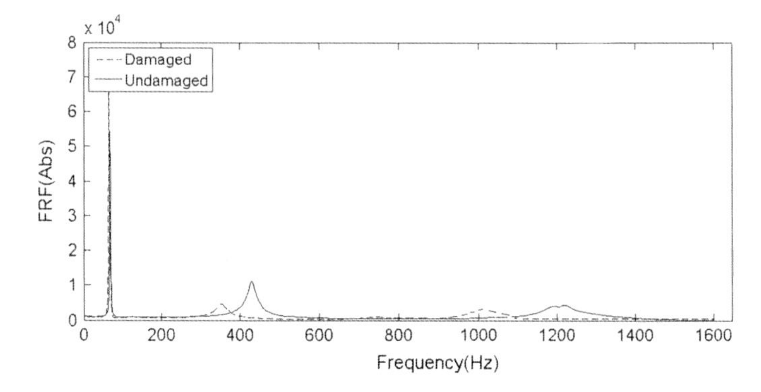

Fig. 7 Frequency response for delaminated and intact beam obtained by FFT analyzer (3107 B & K)

laser vibrometer (Ometron VH 300) targeted at 47 mm from the fixed end. The response measurements were acquired using signal analyzer (Pulse 2827 B & K) and the frequency response function (FRF) of the beam was calculated using FFT analyzer (3107 B & K). The Polynomial Fit method of ME'scope ™ software was used to extract accurate natural frequencies of the composite beam from extracted FRF [19]. The mentioned FRF of the beam in each case is shown in Fig. 7

8 Conclusion

Advanced methods of supervision, fault detection and fault diagnosis become increasingly important in many engineering systems for the improvement of reliability, safety and efficiency. The proposed methodology in this paper works on frequency domain, and it uses AIS to detect the faults. The components of the AIS such as its representation, affinity function, and immune process are tailored for the damage detection. The evolution and immune learning algorithms make it possible for the damage detector to generate a high quality memory cell set for recognizing various structure damages. The comparison study of the damage detector accuracy with other damage detectors such as ANN and BGA has also been conducted. For the delaminated beam, the AIS show a higher detection success rate when comparing with ANN methods. The VAF value of AIS is greater than those of ANN-based different methods. The prediction errors of AIS are less than those of the BGA. The average values of location and length prediction errors are 2.16 and 0.1968 % for the AIS method, respectively which those values become −3.83 and 1.76 % for the BGA . The effectiveness of this method is approved by experimental results, the prediction values for location and length errors are 18.7 and 9.8 %, respectively. The proposed model has the ability to be used for analysis of complex structures with different kinds of damages.

References

1. Chakraborty, D.: Artificial neural network based delamination prediction in laminated composites. Mater. Des. **26**, 17 (2005)
2. Valoor, M.T., Chandrashekhara, K.: A thick composite-beam model for delamination prediction by the use of neural networks. Compos. Sci. Technol. **60**, 1773–1779 (2000)
3. Garg, A.C.: Delamination a damage mode in composite structures. Engrg. Frac. Mech. **29**, 557–584 (1988)
4. Tay, T.E.: Characterization and analysis of delamination fracture in composites: an overview of developments from 1990 to 2001. Appl. Mech. Rev. **56**, 131 (2003)
5. Della, C.N., Shu, D.: Free vibration analysis of composite beams with overlapping delaminations. Eur. J. Mech. A/Solids **24**, 491–503 (2005)
6. Doebling, S.W., Farrar, C.R., Prime, M.B., Shevitz, D.W.: Damage identification and health monitoring of structural and mechanical systems from changes in their vibration characteristics: A Literature Review, Los Alamos National Laboratory Report, May 1996; LA-13070-MS
7. Dasgupta, D: Advances in artificial immune systems. Comput. Intell. Mag. IEEE. **1**, 40–49 (2006)
8. Timmis, J., Hone, A.N.W., Stibor, T., Clark, E.: Theoretical advances in artificial immune systems. Theoret. Comput. Sci. **403**, 11–32 (2008)
9. Timmis, J.: Application areas of AIS: the past, the present and the future. Appl. Soft Comput. **3**, 191–201 (2008)
10. Chen, B., Zang, C.H.: Artificial immune pattern recognition for structure damage classification. Comput. Struct. **87**, 1394–1407 (2009)
11. Della, C.N., Shu, D.: Vibration of delaminated composite laminates: a review, Appl. Mech. Rev. **60**, 120 (2007)
12. Shu, D., Della, C.N.: Free vibration analysis of composite beams with two non-overlapping delaminations. Int. J. Mech. Sci. **46**, 509–526 (2004)
13. Mujumdar, P.M., Suryanarayan, S.: Flexural vibrations of beams with delaminations. J. Sound Vib. **125**, 441–461 (1982)
14. Reddy, J.N.: Mechanics of Laminated Composite Plates, CRC Press, Boca Raton, FL (1997)
15. Shen, M.-H.H., Grady, J.E.: Free vibrations of delaminated beams. AIAA J. **30**, 1361–1370 (1992)
16. Luo, H., Hanagud, S.: Dynamics of delaminated beams. Int. J. Solids Struct. **37**, 1501–1519 (2000)
17. Martins, J., Tomas, P., Sousa, L.: Neural code metrics: analysis and application to the assessment of neural models. Neurocomputing **72**, 2337–2350 (2009)
18. Hein, H., Feklistova, L.: Computationally efficient delamination detection in composite beams using Haar wavelets. Mech. Syst. Sig. Proces. **25**, 2257–2270 (2011)
19. Brüel, Kjr, Data Sheet for MEscope Modal and Structural Analysis, http://www.bksv.com

On Active Vibrations Control of a Flexible Rotor Running in Flexibly-Mounted Journal Bearings

Mohamed M. Eimadany

Abstract Increasing demands on rotating machinery in terms of higher running speed, less weight and noise, safety and longevity require well controlled dynamics of the system. This paper addresses a means of actively controlling the synchronous vibrations of a flexible rotor running in flexibly-mounted journal bearings. An isotropic optimal controller of the anisotropic rotor-bearing system in complex state space is designed. The isotropic controller essentially eliminates the backward unbalance response component, leading to circular whirling. Simulation results are presented which demonstrate that the isotropic optimal control is more efficient in controlling unbalance whirl than the conventional optimal control.

1 Introduction

Vibration is a problem of perpetual concern for rotating machinery manufacturers due to its central role in machinery performance, safety, and reliability. With the rotor machinery customers adopting increasingly stringent requirements for acceptable vibration levels, development of effective vibration reduction strategies has assumed greater importance in recent years. Such mechanisms for dissipating vibration energy include seal dampers [1], squeeze-film dampers [2], hybrid squeeze-film dampers [3], hydraulic active chamber systems [4, 5], variable impedance hydrodynamic journal bearings [6], actively lubricated tilting-pad bearings [7–10], active-controlled hydrostatic bearings [11], magnetized journal bearings lubricated with Ferro fluids [12], and actively lubricated hybrid multirecess journal bearings [13].

Fortunately, new active control techniques now being developed appear capable of effectively reducing the vibration levels. The advantages gained from implementing active vibration control include the capability of being tuned or adjusted and made

M. M. Eimadany(✉)
Mechanical Engineering Department, King Saud University, Riyadh, Saudi Arabia
e-mail: mmadany@ksu.edu.sa

P. A. Muñoz-Rojas (ed.), *Optimization of Structures and Components*,
Advanced Structured Materials 43, DOI: 10.1007/978-3-319-00717-5_6,

self adaptive to varying conditions of loading and rotor speed, the use of more general control forces in comparison to those available from spring and damper elements, and the attenuation of the vibration amplitude during run-up and coast down through the critical speeds.

Numerous studies have been carried out on active control of rotating machinery for safe operation. Zhu et al. [14] and Abduljabbar et al. [15] have studied a control law which accommodates unbalance force and feedback of estimated state and unbalance using observer. The periodic learning control with inverse transfer function compensation in magnetic bearing system for suppressing the unbalance response has been investigated by Higuchi et al. [16]. The application and robustness of feedforward compensater to suppress the synchronous vibration of rotor systems has been investigated by Knospe [17], Larsonneur et al. [18], and Abduljabbar et al. [19]. Mizutani et al. [20] used a semi-active vibration control system to reduce the unbalanced vibration of a rotor. In a paper by Eimadany and Abduljabbar [21], the synthesis of an optimal control law using linear quadratic regulator theory accompanied by an asymptotic state observer has been performed. They have shown that the control law has the compatibility to attenuate the lateral vibration and stabilize a rotor system with anisotropic fluid-film bearings and fluid leakage. Li et al. [22] applied active piezoceramic pusher bearings and self-learning control for the suppression of the rotor system vibration. Nicoletti [23] presented a methodology based on modal reduction for calculating the gains of an output feedback controller for active vibration control of flexible rotors. Tammi [24] examined repetitive learning control for active vibration control of rotor. Ondrouch et al. [25] used feedback controllers for active reduction of lateral vibration of symmetric, rigid rotor supported by journal bearings. Juhanko et al. [26] presented an industrial application of the active control of bearing o a paper machine roll. Tüma et al. [27] investigated experimentally the active control of journal bearing. An overview of the theoretical and experimental achievements of mechatronics applied to fluid film bearings was presented by Santos [28]. He explored the ideas of combining control techniques, informatic with hyrdodynamic, thermo-hydraulic, elastrodynamic and thrmo-elaso-hydrodynamic lubrication techniques.

Using complex notation, Lee and Byun [29], and Kim and Lee [30] have studied the optimal control of isotropic rotor bearing system using complex state space. Kim and Lee [31] presented an isotropic optimal control scheme for weakly anisotropic rotor bearing system. Isotropic optimal controller design in complex state space is essentially composed of two steps. In the first step, the system is decomposed into isotropic and anisotropic parts, and direct canceling control of the system anisotropy is carried out. In the second step, an isotropic optimal control scheme is applied to the resulting isotropic system.

In this paper, an isotropic control law for a rotor system with anisotropic fluid-film bearings is derived. The performance of the isotropic optimal control is compared with that of the conventional optimal control method.

2 Rotor-Bearing System Model

Figure 1 shows the simplified model to investigate the vibration control of a rotor-bearing system mounted on a damped flexible support. This model consists of a rotor symmetrically mounted on two fluid-film bearings supported with damped flexible bearings. The rotor is modelled as three lumped masses and connected by massless flexible shafts. Control forces are exerted on the bearings in both vertical and horizontal directions.

Under the assumption of small vibration, the nonlinear characteristics of the fluid-film bearing can be linearized at the static equilibrium position. The dynamic characteristics of the fluid-film bearing are represented by eight stiffness and damping

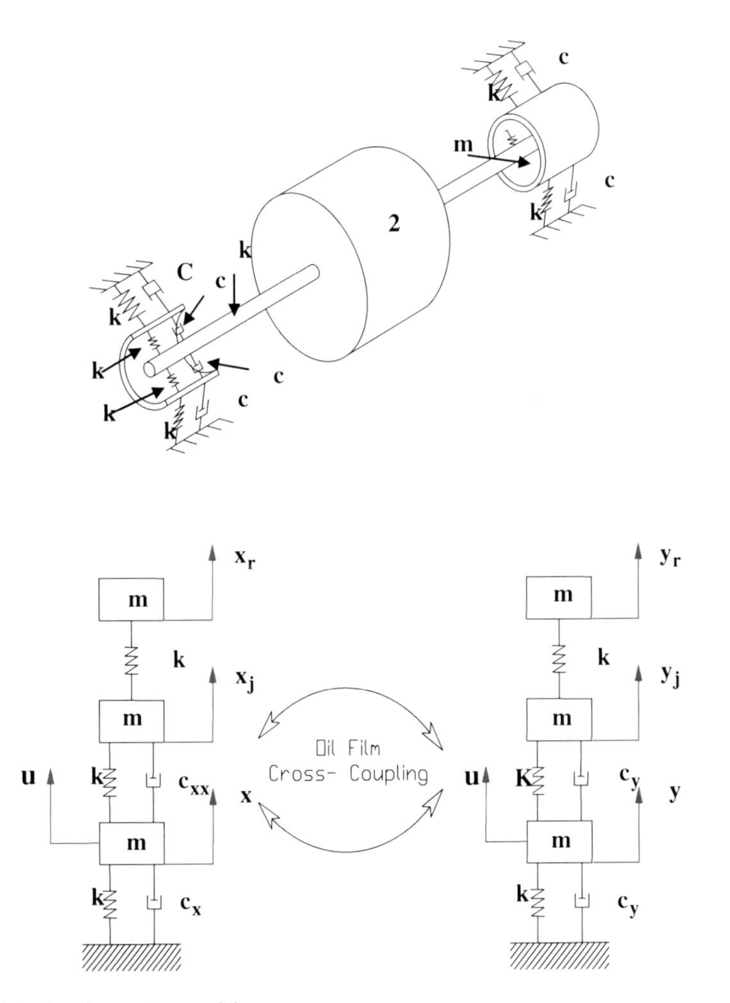

Fig. 1 Rotor-bearing system model

coefficients, k_{ij} and c_{ij} (i, j = x, y). These coefficients vary with the rotational speed and significantly influence the vibration response amplitude and stability.

The forces acting on the journal and bearing can be expressed as (see Fig. 1):

$$\begin{bmatrix} F_{xj} \\ F_{yj} \end{bmatrix} = -\begin{bmatrix} c_{xx} & c_{xy} \\ c_{yx} & c_{yy} \end{bmatrix}\left[\begin{pmatrix} \dot{x}_j - \dot{x}_b \\ \dot{y}_j - \dot{y}_b \end{pmatrix}\right] - \begin{bmatrix} k_{xx} & k_{xy} \\ k_{yx} & k_{yy} \end{bmatrix}\left[\begin{pmatrix} x_j - x_b \\ y_j - y_b \end{pmatrix}\right] \tag{1}$$

$$\begin{aligned}\begin{bmatrix} F_{xb} \\ F_{yb} \end{bmatrix} = &-\begin{bmatrix} c_{xx} c_{xy} \\ c_{yx} c_{yy} \end{bmatrix}\left[\begin{pmatrix} \dot{x}_b - \dot{x}_j \\ \dot{y}_b - \dot{y}_j \end{pmatrix}\right] - \begin{bmatrix} k_{xx} k_{xy} \\ k_{yx} k_{yy} \end{bmatrix}\left[\begin{pmatrix} x_b - x_j \\ y_b - y_j \end{pmatrix}\right] \\ &- \begin{bmatrix} c_x 0 \\ 0 c_y \end{bmatrix}\begin{bmatrix} \dot{x}_b \\ \dot{y}_b \end{bmatrix} - \begin{bmatrix} k_x 0 \\ 0 k_y \end{bmatrix}\begin{bmatrix} x_b \\ y_b \end{bmatrix}\end{aligned} \tag{2}$$

where the subscript j refers to the journal and the subscript b refers to the bearing.

The equations of motion of the multi-degree-of-freedom rotor bearing system shown in Fig. 1, may be written in matrix form as [15, 21]

$$M\ddot{q} + C\dot{q} + Kq = f_e + f_c \tag{3}$$

where

$$q = \begin{Bmatrix} x \\ y \end{Bmatrix}, \; f_e = \begin{Bmatrix} f_{ex} \\ f_{ey} \end{Bmatrix}, \; f_c = \begin{Bmatrix} f_{cx} \\ f_{cy} \end{Bmatrix}, \; M = \begin{bmatrix} M_{xx} & M_{xy} \\ M_{yx} & M_{yy} \end{bmatrix},$$

$$C = \begin{bmatrix} C_{xx} & C_{xy} \\ C_{yx} & C_{yy} \end{bmatrix}, \; K = \begin{bmatrix} K_{xx} & K_{xy} \\ K_{yx} & K_{yy} \end{bmatrix} \tag{4}$$

Here M, C, and K are mass, damping and stiffness matrices of appropriate dimensions, x(y), $f_{ex}(f_{ey})$, $f_{cx}(f_{cy})$ are x− (y−) directional displacement, excitation force, and control force vectors, respectively.

$$M_{xx} = M_{yy} = \begin{bmatrix} m_r & 0 & 0 \\ 0 & m_j & 0 \\ 0 & 0 & m_b \end{bmatrix}, \quad M_{xy} = M_{yx} = 0$$

$$C_{xx} = \begin{bmatrix} c_s & -c_s & 0 \\ -c_s & c_s + c_{xx} & -c_{xx} \\ 0 & -c_{xx} & c_{xx} + c_x \end{bmatrix}, \quad C_{yy} = \begin{bmatrix} c_s & -c_s & 0 \\ -c_s & c_s + c_{yy} & -c_{yy} \\ 0 & -c_{yy} & c_{yy} + c_y \end{bmatrix}$$

$$C_{xy} = \begin{bmatrix} 0 & 0 & 0 \\ 0 & c_{xy} & -c_{xy} \\ 0 & -c_{xy} & c_{xy} \end{bmatrix}, \quad C_{yx} = \begin{bmatrix} 0 & 0 & 0 \\ 0 & c_{yx} & -c_{yx} \\ 0 & -c_{yx} & c_{yx} \end{bmatrix}$$

$$K_{xx}=\begin{bmatrix}k_s & -k_s & 0\\ -k_s & k_{ks}+k_{xx} & -k_{xx}\\ 0 & -k_{xx} & k_{xx}+k_x\end{bmatrix},\quad K_{yy}=\begin{bmatrix}k_s & -k_s & 0\\ -k_s & k_s+k_{yy} & -k_{yy}\\ 0 & -k_{yy} & k_{yy}+k_y\end{bmatrix}$$

$$K_{xy}=\begin{bmatrix}0 & 0 & 0\\ 0 & k_{xy} & -k_{xy}\\ 0 & -k_{xy} & k_{xy}\end{bmatrix},\quad K_{yx}=\begin{bmatrix}0 & 0 & 0\\ 0 & k_{yx} & -k_{yx}\\ 0 & -k_{yx} & k_{yx}\end{bmatrix}$$

$$f_{ex}=\begin{bmatrix}m_r a\Omega^2\cos\Omega t\\ 0\\ 0\end{bmatrix},\quad f_{ey}=\begin{bmatrix}m_r a\Omega^2\sin\Omega t\\ 0\\ 0\end{bmatrix}$$

$$f_{cx}=\begin{bmatrix}0\\ 0\\ 1\end{bmatrix}u_1,\quad f_{cy}=\begin{bmatrix}0\\ 0\\ 1\end{bmatrix}u_2 \tag{5}$$

The system is decomposed into isotropic and anisotropic parts.

Let

$$p=x+jy,\ \bar{p}=x-jy,\ g=f_{ex}+jf_{ey},\quad h=f_{cx}+jf_{cy} \tag{6}$$

Then

$$M_i\ddot{p}+C_i\dot{p}+K_ip+C_a\dot{\bar{p}}+K_a\bar{p}=g+h \tag{7}$$

where

$$M_i=M_{xx}=M_{yy}$$

$$K_i=\frac{K_{xx}+K_{yy}}{2}+j\frac{K_{yx}-K_{xy}}{2},\quad K_a=\frac{K_{xx}-K_{yy}}{2}+j\frac{K_{yx}+K_{xy}}{2}$$

$$C_i=\frac{C_{xx}+C_{yy}}{2}+j\frac{C_{yx}-C_{xy}}{2},\quad C_a=\frac{C_{xx}-C_{yy}}{2}+j\frac{C_{yx}+C_{xy}}{2} \tag{8}$$

Here, M_i, C_i, K_i represent the isotropic properties of the rotor bearing system. C_a, K_a represent the anisotropic properties of bearings. $g=\hat{g}e^{j\Omega t}$, and '–' denotes the complex conjugate.

2.1 Unbalance Response

Let the solution be

$$p=p_f\,e^{j\Omega t}+p_b\,e^{-j\Omega t} \tag{9}$$

where p_f and p_b are the forward and backward whirl response vectors. The unbalance response can be obtained from

$$\begin{bmatrix} H_{ff} & H_{fb} \\ H_{bf} & H_{bb} \end{bmatrix} \begin{bmatrix} p_f \\ p_b \end{bmatrix} = \begin{bmatrix} \hat{g} \\ 0 \end{bmatrix} \tag{10}$$

where

$$H_{ff} = -\Omega^2 M_i + j\Omega C_i + K_i, \quad H_{fb} = j\Omega C_a + K_a$$
$$H_{bb} = -\Omega^2 M_i + j\Omega \bar{C}_i + \bar{K}_i, \quad H_{bf} = j\Omega \bar{C}_a + \bar{K}_a$$

From Eq. (10), it is clear that $p_b = 0$ if $H_{bf} = H_{fb} = 0$, or:

$$j\Omega\, C_a + K_a = 0 \tag{11}$$

In this case, the response is represented by forward synchronous whirl as

$$p = p_f\, e^{j\Omega t} = H_{ff}^{-1}\, \hat{g e}^{j\Omega t} \tag{12}$$

3 Conventional Optimal Control

The state space form of the rotor bearing system 3 can be written as

$$\dot{x} = Ax + Bu \tag{13}$$

where A is the system matrix, B is the control input matrix and u is the control input.

$$A = \begin{bmatrix} 0 & I \\ -M^{-1}K & -M^{-1}C \end{bmatrix}, \quad B = \begin{bmatrix} 0 \\ M^{-1} \end{bmatrix}, \quad x = \begin{bmatrix} q \\ \dot{q} \end{bmatrix}, \quad u = f_c$$

For a linear time-invariant system 13, a full optimal state feedback control law may be synthesized by minimizing the exponentially weighted performance index

$$J = \int_0^\infty e^{2\alpha t} \left(x^T Q x + u^T R u \right) dt \tag{14}$$

where R and Q are constant, symmetric matrices with R positive definite and Q positive semi-definite. α is a non-negative constant introduced to provide a minimum degree of stability for the closed-loop system. The optimal control law which minimizes the performance index 14 is

$$u = -Kx = -R^{-1}B^T Px \tag{15}$$

where P is the symmetric, positive definite matrix solution of the steady-state Riccati equation

$$\left(A^T + \alpha I\right) P + P\,(A + \alpha I) - PBR^{-1}B^T P + Q = 0 \tag{16}$$

The optimal control gain, K, and the control force vector can be written as

$$K = \left[K_p\, K_d\right], \quad u = \left[K_p q + K_d \dot{q}\right] \tag{17}$$

where K_p and K_d are the $2n \times 2n$ proportional and derivative gain matrices. Equation 3 can be written as

$$M\ddot{q} + [C + K_d]\,\dot{q} + \left[K + K_p\right] q = f_e \tag{18}$$

or in complex form

$$M_i\ddot{p} + [C_i + K_{di}]\,\dot{p} + \left[K_i + K_{pi}\right] p + [C_a + K_{da}]\,\dot{\bar{p}} + \left[K_a + K_{pa}\right]\bar{p} = g \tag{19}$$

In general, conventional optimal controlled rotor bearing system retains the characteristics of original system. Therefore, if the original system is anisotropic, the controlled system is also likely to be anisotropic.

4 Isotropic Optimal Control

Let the control force be h and is given by:

$$h = h_i + h_a \tag{20}$$

where

$$h_a = C_a\dot{\bar{p}} + K_a\bar{p} \tag{21}$$

Consider the isotropic system

$$M_i\ddot{p} + C_i\dot{p} + K_i p = h_i \tag{22}$$

or

$$\dot{x}_i = A_i\, x_i + B_i\, u_i \tag{23}$$

where

$$x_i = \begin{bmatrix} p \\ \dot{p} \end{bmatrix},$$

$$A_i = \begin{bmatrix} 0 & I \\ -M_i^{-1}K_i & -M_i^{-1}C_i \end{bmatrix}, \quad B_i = \begin{bmatrix} 0 \\ M_i^{-1} \end{bmatrix}, \quad u_i = h_i$$

Let the performance index J_i given by

$$J_i = \int_0^{\infty} e^{2\alpha t} \left(x_i^* Q_i x_i + u_i^* R_i u_i \right) dt \tag{24}$$

where Q_i and R_i are the positive semi-definite and positive definite Hermitian matrices, respectively, and * the conjugate transpose. Then the solutions to the minimization of J_i is the complex optimal control law given by

$$u_i = -R_i^{-1} B_i^* P_i x_i = -\left[K_{p_i}\ K_{d_i}\right] x_i \tag{25}$$

where the positive definite Hermitian matrix P_i is the solution of the complex valued algebraic Riccati equation:

$$(A_i^* + \alpha I) P_i + P_i (A_i^* + \alpha I) - P_i B_i R_i^{-1} B_i^* P_i + Q_i = 0 \tag{26}$$

The complex control force vector becomes, using feedback gain matrices

$$u_i = -\left[K_{pi} p + K_{di} \dot{p}\right] \tag{27}$$

and the final control law is the superposition of the two control actions given in Eqs. 21 and 27, i.e.,

$$\begin{aligned} u &= u_i + h_a \\ &= -\left[K_{pi} p + K_{di} \dot{p}\right] + \left[K_a \bar{p} + C_a \dot{\bar{p}}\right] \end{aligned} \tag{28}$$

The controlled system becomes then

$$M_i \ddot{p} + [C_i + K_{di}]\, \dot{p} + \left[K_i + K_{pi}\right] p = g \tag{29}$$

5 Simulation Results and Discussion

The simulation is based on the following system parameters: Rotor mass, m_r = 306.06 kg, Journal mass, m_j = 141, 47 kg (each), Bearing mass, m_b = 6.80 kg (each), Shaft stiffness, $k_s = 49 \times 10^6$ N/m, Shaft damping, $c_s = 0$, Support stiffness, $k_x, k_y = 10 \times 10^6$ N/m, Support damping $c_x, c_y = 5000$ Ns/m.

The bearing stiffness and damping coefficients are calculated using short bearing model [34, 34]. The bearing parameters are: Radius, r = 0.0305 m, Land length, λ = 0.0254 m, Radial clearance, c = 0.000102 m, Lubricant viscosity, $\mu = 0.069$ Ns/m^2.

Table 1 Open-loop eigenvalues of the rotor-bearing model

Definition	Eigenvalues at rotor speed of	
	170 rad/s	830 rad/s
Fluid-film modes	-21.9×10^4, $-1.77 \times 10^4 - 240 \pm j153$	-4.1×10^4, -0.83×10^4 $-609 \pm j524$
Flexible rotor modes	$-9.5 \pm j827$ $-9.5 \pm j828$	$-8.4 \pm j827 - 9.5 \pm j828$
Suspension modes	$-9.6 \pm j157$ $-8.7 \pm j175$	$-2.9 \pm j166 - 10 \pm j172$

Table 2 Closed-loop eigenvalues at 830 rad/s

Conventional optimal control	Isotropic optimal control
-4.8×10^4, -0.8×10^4	$-2.5 \times 10^4 + j403$
$-599 \pm j507$	$-547 - j392$
$-92 \pm j826$	$-106.5 + j824$
$-86 \pm j828$	$-104 - j827$
$-87 \pm j163$	$-105 + j162$
$-90 \pm j169$	$-97 - j169$

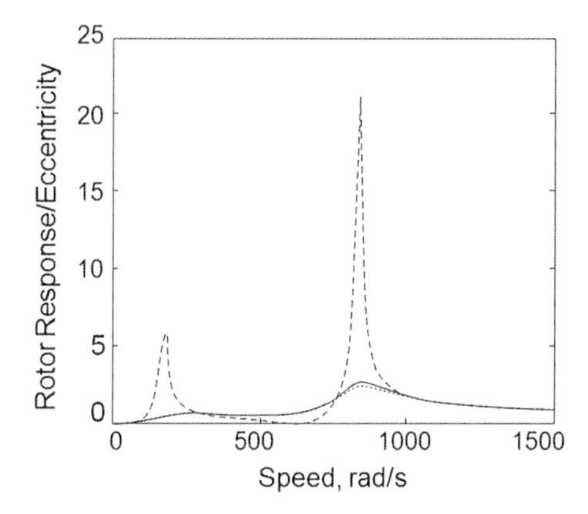

Fig. 2 Steady-state rotor response, (*Dashed with space lines* uncontrolled, (*Dashed lines*) conventional, (*Dotted lines* isotropic

The open-loop eigenvalues of the twelfth-order model of the rotor-bearing system at running speeds of 170 rad/s and 830 rad/s are listed in Table 1. The open-loop rotor exhibits marginally stable suspension modes with frequency of approximately 175 rad/s. The rotor modes are also marginally stable with frequencies of approximately 827 rad/s. It is clear that the rotor-bearing system at hand has two critical speeds at approximately 175 rad/s and 828 rad/s with very little damping in the corresponding whirl modes. The state weighting matrices for the closed-loop are chosen so as to satisfy several design criteria. In each case, α, Q, Q_i, R, and R_i are chosen to position the closed-loop system poles such that they provide acceptable transient response behavior, and reasonable system bandwidth and gains. This is done based on a method suggested in [34]. The eigenvalues at a running speed of 830 rad/s for the conventional optimal controller and isotropic optimal con-

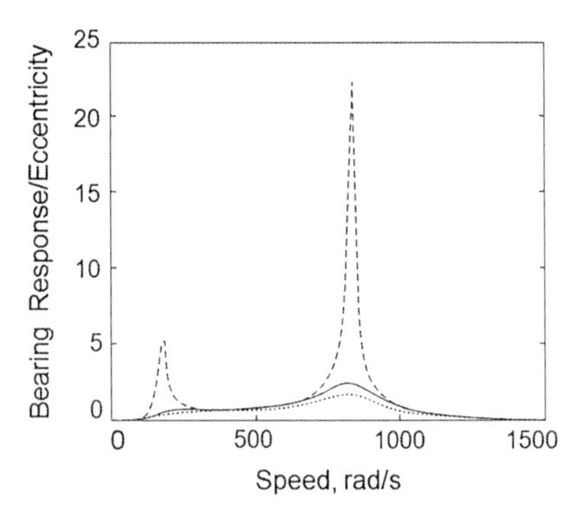

Fig. 3 Steady-state bearing response, (*Dashed with space lines*) uncontrolled, (*Dashed lines*) conventional, (*Dotted lines*) isotropic

troller are shown in Table 2. The calculations of the control gains are based on $\alpha = 40, \mathrm{Q} = \mathrm{I}_{12}, \mathrm{Q_i} = \mathrm{I}_6, \mathrm{R} = \mathrm{R_i} = 0.5 \times 10^8 \mathrm{I}_2$, where I_n is an identity matrix of order nxn.

Table 2 shows that both the conventional and isotropic optimal controlled systems have similar eigenvalues, particularly for flexible rotor modes and suspension modes but with more damping near the second critical speed.

Typical steady-state responses of the actively controlled rotor-bearing system using conventional and isotropic optimal controllers are shown in Figs. 2 and 3, together with the results of the uncontrolled system. The response amplitudes are based on the major axis of the whirling motions. In isotropic optimal controlled rotor-bearing system, unbalance responses are theoretically characterized by a forward synchronous circular whirl. It is clear that the major whirl radius is smaller for the isotropic optimal control than that for the conventional optimal control.

6 Concluding Remarks

Rotor-bearing systems have peculiar modal characteristics known as the forward and backward modes, and the use of complex notation in the analysis enables the separation of these modes. In this paper, an isotropic optimal control of a flexible rotor supported on anisotropic flexibly-mounted fluid-film bearings in complex state space is designed. The control effort of the isotropic optimal control is twofold. The first part of the control action is devoted to make the system isotropic by direct canceling control of system anisotropy. The second part of control action is applied to the resulting isotropic system. For the synthesis of the controller, linear quadratic regulator technique in complex state space is used to determine the feedback gains. The resulting controlled system retains isotropic eigenstructure; leading to circular whirling due to mass unbalance.

From the simulation results, it can be concluded that the isotropic optimal control method is more effective in suppressing the unbalance vibrations than the conventional optimal control.

Acknowledgments The author would like to thank the Research Center, College of Engineering, King Saud University for supporting this work under grant number 5/429 from SABIC. The assistance and encouragement of the Research Center are very much appreciated.

References

1. Vance, J.M., Li, J.: Test results of a new damper seal for vibration reduction in turbomachinery. ASME J. Eng. Gas Turbines Power **118**(4), 843–846 (1996)
2. San Andrés, L., Lubell, D.: Imbalance response of a test rotor supported on squeeze film dampers. ASME J. Eng. Gas Turbines Power **120**(2), 397–404 (1998)
3. El-Shafei, A., Dimitri, A.S.: Controlling journal bearing instability using active magnetic bearings. In: Proceedings of ASME Turbo Expo : Paper GT2007-28059. Montreal, Canada (2007)
4. Ulbrich, H., Althaus, J.: Actuator Design for Rotor Control. In: proceedings, 1989, 12th Biennial ASME Conference on Vibration and Noise, Montreal, Sept.17-21, pp. 17–22 (1989)
5. Santos, I.F.: On the adjusting of the dynamic coefficients of tilting-pad journal bearings. STLE Tribol. Trans. **38**(3), 700–706 (1995)
6. Goodwin, M.J., Boroomand, T., Hooke, C.J.: Variable impedance hydrodynamic journal bearings for controlling flexible rotor vibrations. In: Proceedings, 1989, 12th Biennial ASME Conference on Vibration and Noise, Montreal, 17–21 Sept. pp. 261–267 (1989)
7. Santos, I.F.: Design and evaluation of two types of active tilting-pad journal bearings. In: Proceedings, pp. 79–87. IUTAM Symposium on Active Control of Vibration, Bath, England (1994)
8. Santos, I.F., Russo, F.H.: Tilting-pad journal bearings with electronic radial oil injection. ASME J. Tribol. **120**(3), 583–594 (1998)
9. Santos, I.F., Nicoletti, R.: THD analysis in tilting-pad journal bearings using multiple orifice hybrid lubrication. ASME Trans. J. Tribol. **121**(4), 892–900 (1999)
10. Santos, I.F., Scalabrin, A.: Control system design for active lubrication with theoretical and experimental examples. ASME J. Eng. Gas Turbine Power **125**(1), 75–80 (2003)
11. Bently, D.E., Grant, J.W., Hanifan, P.C.: Active controlled hydrostatic bearings for a new generation of machines. In: proceedings, 2000, ASME/IGTI International Gas Turbine & Aeroengine Congress & Exhibition, Munich, Germany, may 8–11, Paper 2000-GT-354
12. Osman, T.A., Nada, G.S., Safar, Z.S.: Static and dynamic characteristics of magnetized journal bearings lubricated with ferrofluid. Tribol. Int. **34**(6), 369–380 (2001)
13. Santos, I.F., Watanabe, F.Y.: Feasibility of influencing the dynamic fluid film coefficients of a multirecess journal bearing by means of active hybrid lubrication. RBCM—J. Brazilian Soc. Mech. Sci. **25**(2), 154–163 (2003)
14. Zhu, W., Castelazo, I., Nelson, H.D.: An active optimal control strategy of rotor vibrations using external forces. In: Proceedings, : ASME Design Technical Conferences—12th Biennial Conference on Mechanical Vibration and Noise. Rotating Machinery Dynamics, DE **18**, 351–359 (1989)
15. Abduljabbar, Z., ElMadany, M.M., Abdulwahab, A.A.: Active Vibration Control of a Flexible Rotor. Comput Struct. **58**(3), 499–511 (1996)
16. Higuchi, T., Otsuka, M., Mizuno, T., Ise, T.: Application of periodic learning control with inverse transfer function compensation in totally active magnetic bearings. In: Proceedings, 1990, 2nd International Symposium on Magnetic Bearings, Tokyo, pp. 257–264

17. Knopse, C.R.: Robustness of unbalance response controllers. In: Proceedings, 1992, 3rd International Symposium on Magnetic Bearings, Alexandria, pp. 580–589
18. Larsonneur, R., Siegwart, R., Traxler, A.: Active magnetic bearings control strategies for solving vibration problems in industrial rotor systems. In: Proceedings, 1992, IMechE Conf. on Vibrations in Rotating, Machinery, C432/088, pp. 83–90
19. Abduljabbar, Z., ElMadany, M., Al-Bahkali, E.: On the vibration and control of a flexible rotor mounted on fluid film bearings. Comput Struct. **65**(6), 849–856 (1997)
20. Mizutari, K., Asai, A., Katok, K.: Semi-active vibration control for overhung rotor system. Trans. Japan Soc. Mech. Eng. Part C **63**(616), 4102–4107 (1997)
21. ElMadany, M.M., Abduljabbar, Z.: Controller design for high-performance turbomachines. J. Vib. Control **6**, 1205–1223 (2000)
22. Li, W., Maisser, P., Enge, H.: Self-learning Control Applied to Vibration Control of a Rotating Spindle by Piezopusher Bearings. In: Proceedings of the Institute of Mechanical Engineers - Part I: Journal of Systems and Control Engineering **218**(3), 185–196 (2004)
23. Nicoletti, R.: Control system design for flexible rotors supported by actively lubricated bearings. J. Vib. Control **14**, 347–374 (2008)
24. Tammi, K.: Gradient-based repetitive learning control for rotor vibration control. Int. J. Intel. Control Sys. **3**, 222–232 (2008)
25. Ondrouch, J., Ferfecki, P., Poruba, Z.: Vibration reduction of rigid rotor supported by journal bearings. Model. Optimization Phys. Sys. **8**, 85–90 (2009)
26. Juhanko, J., Porkka, E., Kuosmanen, P., Valkonen, A., Järviluoma, M.: Active vibration control of a paper machine roll. In: Proceedings, 6th International DAAAM Baltic Conference, 4–26 April 2008, Tallinn, Estonia
27. Tüma, J.. Škuta, J. , Klečka, R. , Los, J., Šimek, J.: A laboratory test stand for active control of journal bearings. In: Proceeding Colloquium Dynamics of Machines 2010, Inst. of Thermomechanics, Prague, 2–3 Feb 2010, pp. 95–100
28. Santos, I.F.: On the future of controllable fluid film bearings. Mech. Ind. **12**, 275–281 (2011)
29. Lee, C.W., Byun, S.W.: Optimal complex modal-space control of rotating disc vibrations. Optimal Control App. Meth. **9**, 357–370 (1988)
30. Kim, C.S., Lee, C.W.: Isotropic control of rotor bearing system. In: Proceedings, 1993, The 14th Biennial ASME Conference on Mechanical Vibration and Noise, Albuquerque, pp. 325–330
31. Kim, C.S., Lee, C.W.: Isotropic optimal control of active magnetic bearing system. J. Dyn. Sys. Meas. Control **118**, 721–726 (1993)
32. Holmes, R.: The vibration of a rigid shaft on short sleeve bearings. J. Mech. Eng. Sci. **2**, 337–341 (1960)
33. Lund, J.W.: Review of concepts of dynamic coefficients for fluid film journal bearings. ASME J. Tribol. **109**, 37–41 (1987)
34. Gawronski, W.K.: Advanced structural dynamics and active control of structures. Springer, New York (2004). ISBN 0387406492

Multi-Disciplinary Constraint Design Optimization Based on Progressive Meta-Model Method for Vehicle Body Structure

S. J. Heo, I. H. Kim, D. O. Kang, W. Y. Ki, S. M. H. Darwish, W. C. Choi and H. J. Yim

Abstract In order to design a vehicle body with high strength and high stiffness, a multi-disciplinary design process should include careful consideration of multi-disciplinary design constraints to properly account for vehicle static stiffness (bending/torsional), durability, Noise/Vibration/Harshness (NVH), crash worthiness, light weight vehicle structure during the early stage of vehicle design process. With this approach, fast development of new vehicle body structures can be achieved with minimal number of iterations to match the conflicting design goals from each discipline. In the current research, a multi-disciplinary design optimization (MDO) based on a meta model is developed and refined to apply for the design of body structure. In an effort to apply the MDO for vehicle body structure, 4 phase procedures were established in the current research. In Phase I, a base model is created. In Phase II, an effect analysis is carried out. In Phase III, a meta model is created. Finally in Phase IV, using the optimization algorithm, the meta model created in Phase III is

S. J. Heo (✉) · I. H. Kim · W. Y. Ki · W. C. Choi · H. J. Yim
Kookim University, Seoul 136702, Korea
e-mail: sjheo@kookmin.ac.kr

I. H. Kim
e-mail: ihkim@kookmin.ac.kr

W. Y. Ki
e-mail: wyki@kookmin.ac.kr

W. C. Choi
e-mail: danchoi@kookmin.ac.kr

H. J. Yim
e-mail: hjyim@kookmin.ac.kr

D. O. Kang
Institute of Design Optimization, Seongnam 463400, Korea
e-mail: bigfive@idopt.co.kr

S. M. H. Darwish
King Saud University, Riyadh 11421, Kingdom of Saudi Arabia
e-mail: darwish@ksu.edu.sa

P. A. Muñoz-Rojas (ed.), *Optimization of Structures and Components*,
Advanced Structured Materials 43, DOI: 10.1007/978-3-319-00717-5_7,

eventually refined through the process of optimization. In this research, static stiffness (bending / torsional), dynamic stiffness (1st torsion mode) were used for constrained conditions and the mass minimization was the object function for optimization.

1 Introduction

As environmental problems are becoming a social issue, importance of green technology is enforced. For instance, lightweight engineering in order to achieve effective fuel consumption normally triggers concerns about high performance and passenger safety from accidental impact.

Thus, these multi-faced design objectives have to be considered simultaneously in a systematic manner during the formalized vehicle body design process in order to build up experiences and technical know-hows for further refinements, especially in the field of lightweight vehicle body design, vehicle safety and structural strength enhancement design. In order to design a vehicle body with high strength and high stiffness, multi-disciplinary contraints should be considered; for instance, static stiffness (bending / torsional), durability performance, noise and vibration, crashworthiness and weight reduction, as seen in Fig. 1 [1–3].

However, since the problem that analysis has to be done repetitively until the differences of the conflicting design schemes among each component are minimized, it is hard to deduct optimal solution which satisfies requirements by using existing opti-

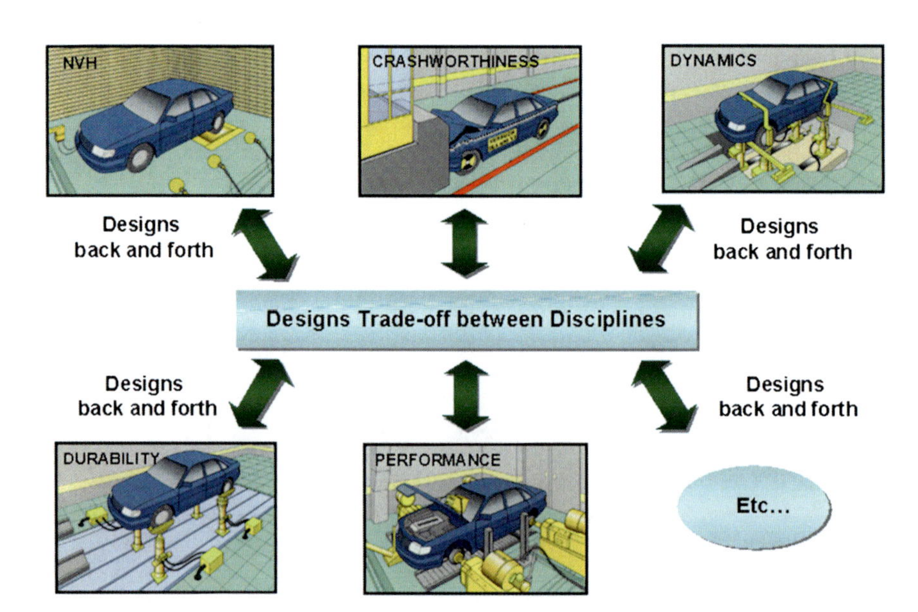

Fig. 1 Current design process of vehicle under various constraints

mal design technology ([4, 5]). Therefore, in the current research, the development of the BIW and its components is greatly facilitated by the use of computational engineering analyses and simulations. Such analyses and simulations are at the core of the multi-disciplinary design optimization (MDO) and design of experiments (DOE) tools, used to support the process of finding "the best" design. In SAE world congress, Anindya Deb compared and evaluated classic response surface method (RSM) and practical MDO methodology. He found out the result that practical MDO approach is cost-effective [6]. In the same conference, Yunkai Gao applied MDO for designing a minivehicle hatchback by using DOE with optimal Latin Hypercube and polynomial RSM [7]. Also, Ford Motor Co. suggested the method of MDO design variable selection by dividing critical parts for design iterations and non-critical for weight reduction [8]. In the present work, a MDO based on a progressive meta model method is used and processes are indicated in Fig. 2

2 FE Model Construction

Within the present work, GAZAL-II which is designed by Saudi Arabia's King Saud University has been considered, as seen in Fig. 3. Static stiffness (bending / torsional) and dynamic stiffness (1st torsion mode) are selected as load cases for FE model construction.

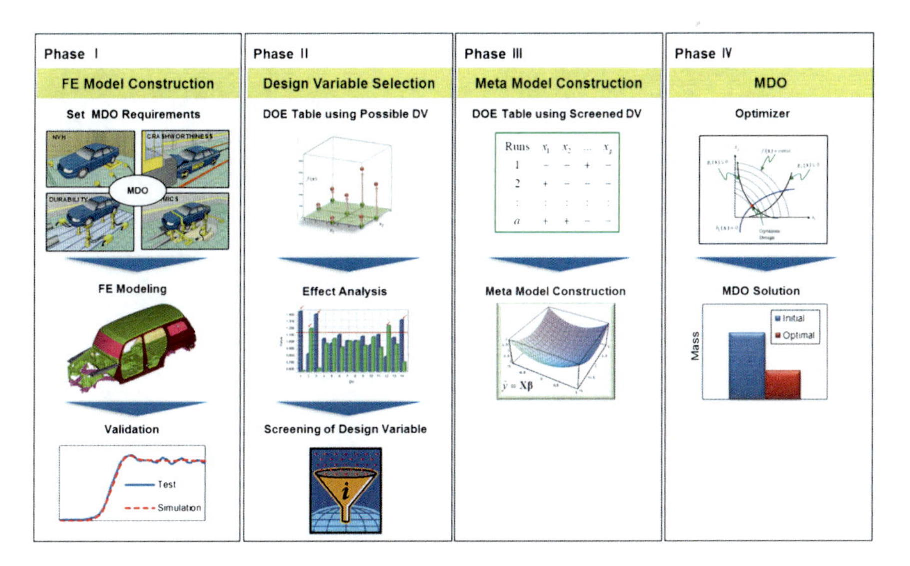

Fig. 2 MDO based on progressive meta model method process

Fig. 3 GAZAL-II

2.1 Static Stiffness (Bending/Torsional)

The static bending stiffness is measured on a static deformation. The boundary condition for the static bending stiffness analysis was implemented with 4 constraints for the front and the rear suspension system (front fixed–dx, dy, dz; rear fixed–dx, dy, dz) and the load is applied at 4 points of passenger compartment as illustrated in

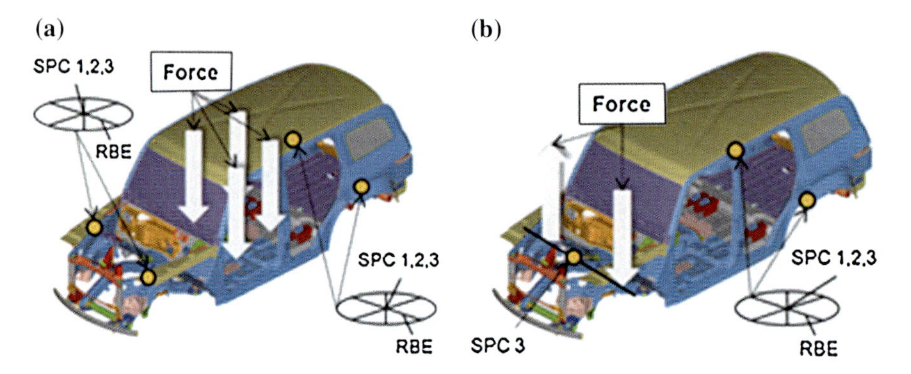

Fig. 4 Boundary condition of static stiffness. **a** Bending analysis. **b** Torsional analysis

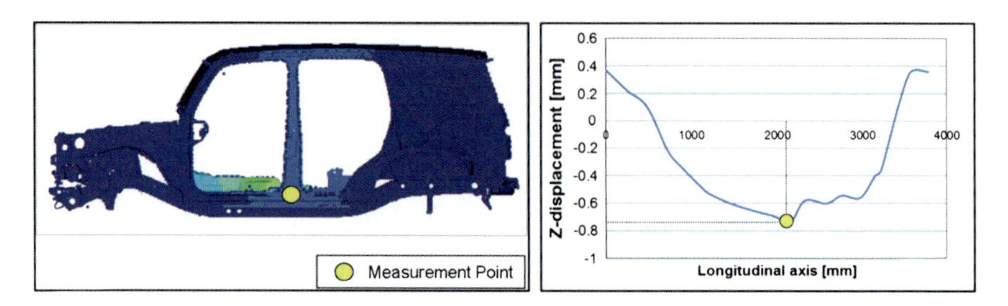

Fig. 5 Result of static stiffness (bending) analysis

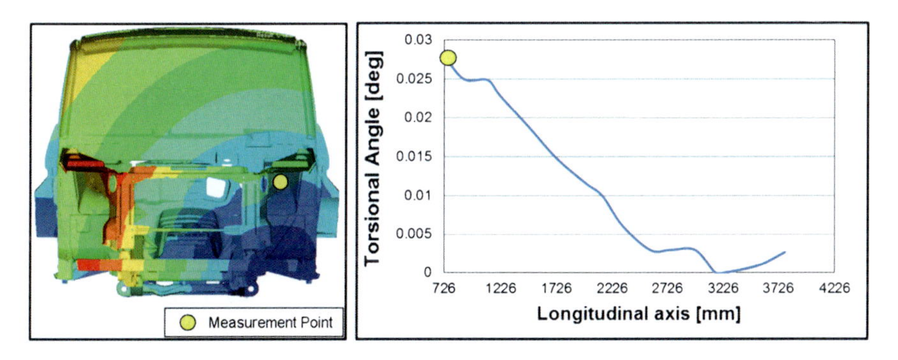

Fig. 6 Result of static stiffness (torsion) analysis

Fig. 4a. The result from the analysis is displayed in Fig. 5. The displacement value is 0.755 mm at B-pillar and rocker panel cross point. The boundary condition for the static torsional stiffness analysis was implemented with 2 constraints for the rear suspension system and another constraint for the front bumper (front fixed–dz; rear fixed–dx, dy, dz) and the torsional loads are applied at 2 points of the front suspension shock mounts, as illustrated in Fig. 4b. To calculate the rotation angle, we selected the middle point of the front/rear shock tower. The result from the analysis is displayed in Fig. 6. The rotation value is 0.028deg at the middle point of the front shock tower. In order to analyze the static stiffness, LS-DYNA CAE tool is used.

2.2 Dynamic Stiffness (1st Torsion Mode)

For a vehicle to be dynamically stiff, it is important to have high natural frequencies for the global modes. In this study, targets are set for these critical global modes of tor-

Fig. 7 Result of dynamic stiffness analysis

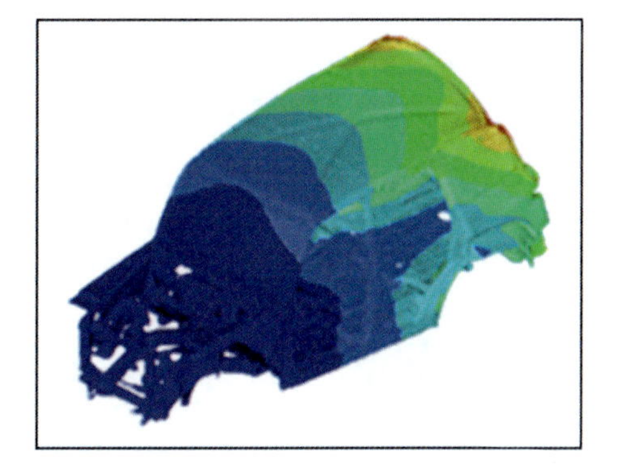

sion that influence the body global stiffness. The result from the analysis is displayed in Fig. 7. In order to analyze the dynamic stiffness, LS-DYNA CAE tool is used.

3 Effect Analysis with Design of Experiments

In the current research, the design variables under consideration for the MDO of the vehicle body structure are about 200. Even with a simple estimation of analysis time itself, it becomes clear that this would be a high cost design process regardless of the accuracy or effectiveness of the optimization. Therefore, it is critical to screen out the design variables that are not significantly affecting the performance indices from various load cases including the bending/torsional stiffness and dynamic stiffness. This process is called "Effect Analysis" and the conceptual schematic diagram of the effect analysis is demonstrated in Fig. 8.

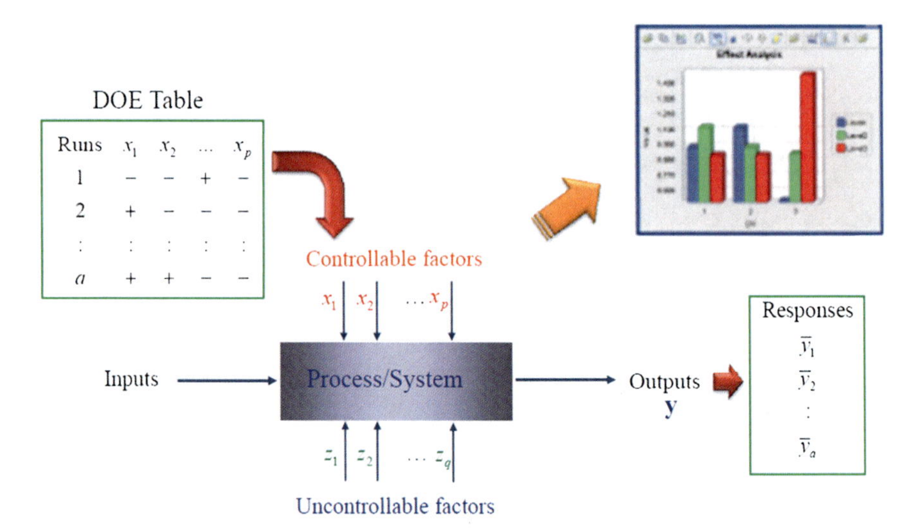

Fig. 8 Effect analysis diagram

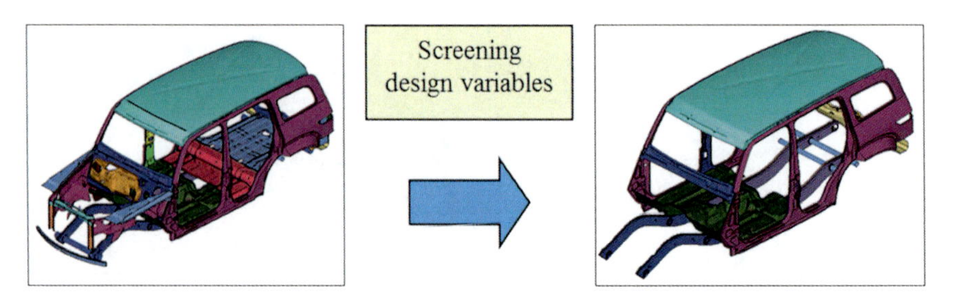

Fig. 9 Initial design variables (45 components) and final design variables (13 components)

In this research, 45 candidate design variables are selected by loadpath which is defined by static/dynamic analysis. And then effect analysis is proceeded to define final design variables. Finally 13 design variables are selected by screening process as shown in Fig. 9. Detail process is described follow chapters.

3.1 Selection of Design Variables for Effect Analysis

First of all, selection of the variables for the effect analysis is required in order to identify the variables affecting static stiffness (bending / torsional), dynamic stiffness (1st torsion mode). Initially, thickness and mass of all parts in the vehicle body under consideration except the front glass and the brackets are selected as design variables.

3.2 Generation of Sampling Points

The total numbers of design variables were identified to be 45. Generation of the sampling points was conducted using a 3-lever ISCD-II (Incomplete Small Composite Design-II) design of experiments provided in EasyDesign, a optimization software. [10] Also, for each design variable, the upper and lower bound of the changes were limited to 20 % from the base model values. The design of experiments for the effect analysis which is generated by EasyDesign is provided in Table 1. For each analysis, the same DOE was used and 48 sampling points were generated.

3.3 Review of Results and Screening Process

Systematic analysis for each load case were carried out for 48 sampling points using LS-DYNA. In Fig. 10, sensitivities of the design variables are graphed for 3 loadcases. Based on this result, the top 13 design variables were screened. It is clear that the frame and side panel affect the bending stiffness. The front rail and the front seat pan

Table 1 Design of experiment table for effect analysis

Run	DV001	DV002	DV003	DV004	...	DV043	DV044	DV045
Run 001	Upper	Lower	Upper	Lower	...	Upper	Lower	Current
Run 002	Lower	Upper	Upper	Upper	...	Current	Lower	Current
Run 003	Lower	Upper	Lower	Upper	...	Upper	Lower	Current
...	...	...	...	...	...	...	...	...
Run 047	Upper	Upper	Upper	Lower	...	Upper	Lower	Current
Run 048	Lower	Lower	Lower	Lower	~	Current	Lower	Current

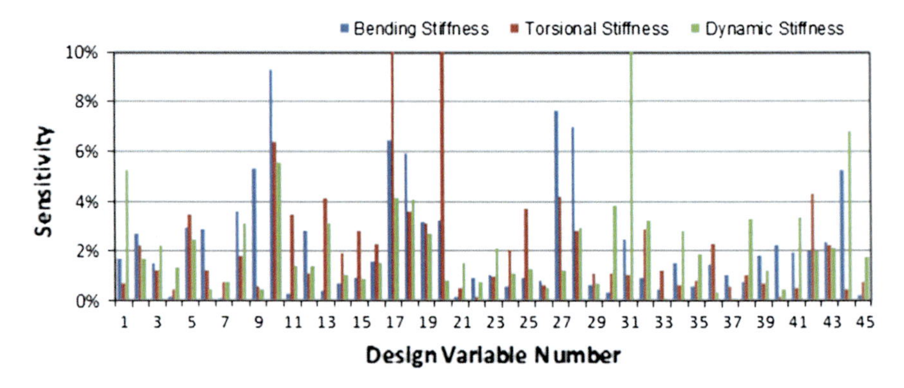

Fig. 10 Screening design variables

affect the torsional stiffness compared to the other design variables. The roof panel, side panel and front rail affect the dynamic stiffness compared to the other design variables.

4 Meta Model Construction and Sequential Approximate Optimization

In order to solve the design optimization problem, a meta model is composed, and optimalization is performed using the composed meta model by a sequential approximate optimization method. LS-DYNA is used as a solver and EasyDesign is used as an optimizor.

4.1 Meta Model Construction

In this section, description for construction of a meta model is provided. First of all, sampling points are created using the design of experiments method for the selected design variables. Then, the response for each sampling point is analyzed to build the meta model that is to be optimized later on. In Fig. 11, a schematic diagram for the process of the construction of the meta model is illustrated.

In this study, radial basis functions (RBF) are used as a meta modeling technique. Radial Basis Function (RBF) is a class of function used for interpolation purposes. Their values depend on only in the distance that is the radius between the generic point and center of the particular function. The RBF method constructs the approximation function $y(x)$ to pass through all sample points using radial basis function $B_i(x)$ and polynomial basis function $X_j(x)$ where w_i is the weighting coefficient for $B_i(x)$ and β_i the coefficient for $X_j(x_i)$. A radial basis function has the following general form:

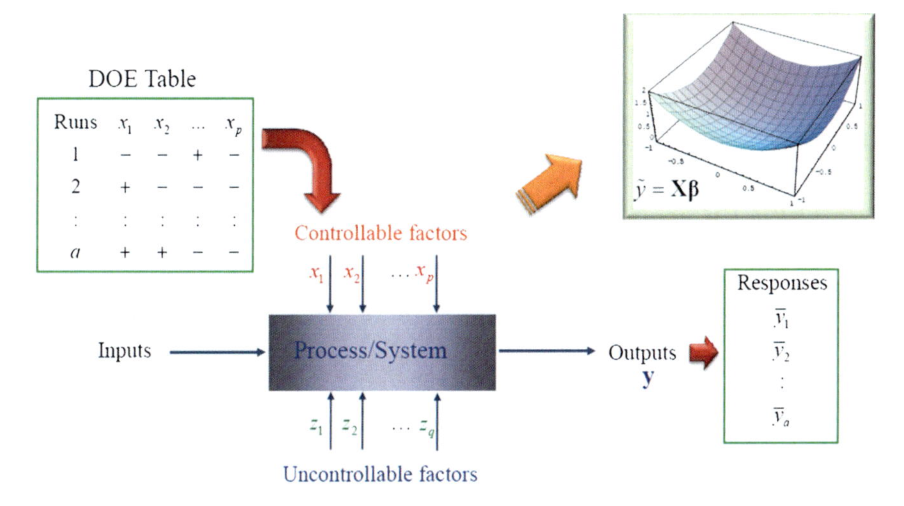

Fig. 11 Meta model construction for optimization

$$y(x) = \sum_{i=1}^{n_s} B_i(x) w_i + \sum_{j=1}^{m} X_j(x_i)\beta_j = \mathrm{B}(x)^T W + \mathrm{X}(x)\beta$$

$$B_i(x) = B_i(r_i)$$

where r_i is the distance between the interpolating point (x) and ith sample point (x_i) [11].

4.2 Sequential Approximate Optimization

The sequential approximate optimization (SAO) process is described in Fig. 12. When the optimization problem is defined, one should select the design variables including random constants and define the design formulation such as objective and constraints. Then the initial sample method is automatically selected to the number of design variables and random constants. When the random constants are given, a Latin hypercube design of 3*k numbers is recommended, where k is the total number of design variables and random constants. When the basic information is given into the SAO process, it solves the user-defined design optimization problem by using numerical optimization algorithms [12]. For constrained optimization problem, an Augmented Lagrange Multiplier (ALM) method is employed. And a quasi-Newton algorithm and a conjugate gradient algorithm are automatically selected to the number of design variables. In order to overcome the divergence due to the lack of sample points, a proper move limit strategy is automatically introduced. Also, the polynomial model is automatically switched to the degree of nonlinearity of responses. When the numerical optimization algorithm is converged in the inner-loop, the

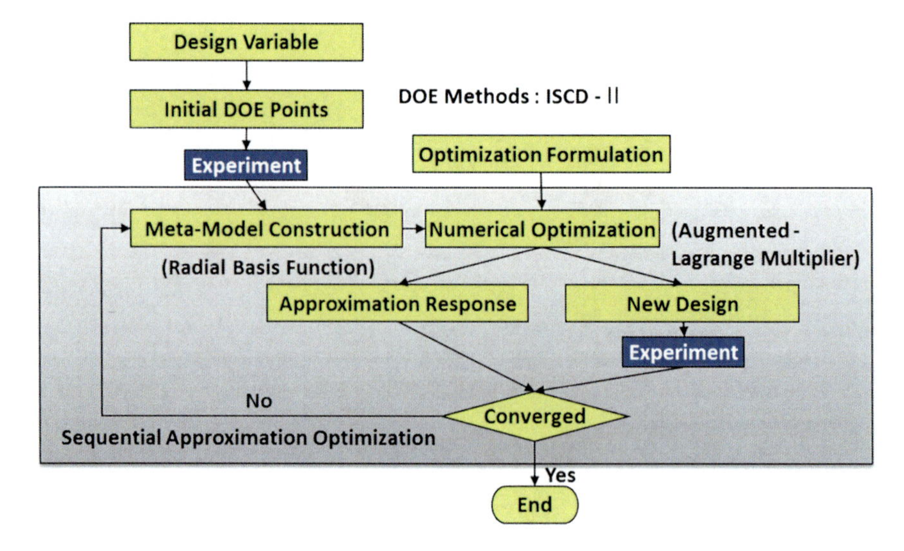

Fig. 12 Process of sequential approximate optimization

convergence is verified through actual analysis results in the outer-loop. At the moment, convergence conditions are not satisfied. Then, this new analysis result is added to the sample results and update a new meta-model and repeat the procedures above. The polynomial type and move limit strategy are automatically selected to the degree of nonlinearity and the magnitude of approximation error.

4.3 Formulation of the Optimization Problem

In this study, a multi-disciplinary design optimization process is formulated to improve static and dynamic stiffness. Accordingly, the objective is to minimize the mass of BIW. The optimization design problem requires satisfying the design constraints (static and dynamic stiffness are made to be 5 % higher than the base model) and minimize the mass of the BIW. This design formulation is mathematically represented below.

$$
\begin{aligned}
&\underset{x_i}{\text{Min}}\,\text{Mass}(x_i)\\
&\text{subject to}\\
&\mathit{Bending\ Stiffness}(x_i) > 5\,\%\ \mathit{higher\ than\ base\ model}\\
&\mathit{Torsional\ Stiffness}(x_i) > 5\,\%\ \mathit{base\ model}\\
&1st\,\mathit{Torsional\ mode}(x_i) > 5\,\%\ \mathit{higher\ base\ model}\\
&X_i^L \le X_i \le X_i^U
\end{aligned}
$$

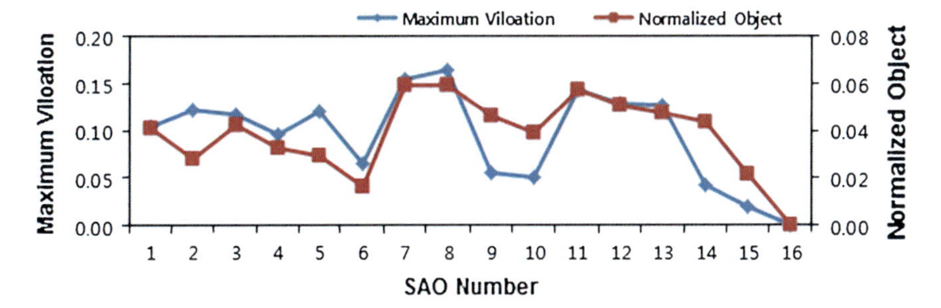

Fig. 13 Convergence history of design optimization

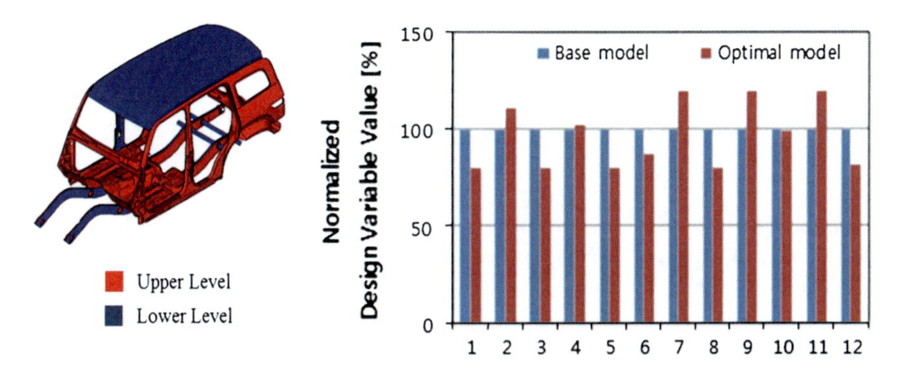

Fig. 14 Comparison of base model and optimal model design variable values

4.4 Results

To solve the design optimization problem, the SAO method based on a meta model is considered. As a result of optimization, mass has been decreased by 0.02 % compared to the existing model, and it is also confirmed that it satisfies the analysis responses selected as constraints. When the approximate optimal solution is calculated by the numerical optimization of the meta model, it is collected, solved and investigated through actual analysis. If the collecting condition is violated, the result of analysis is added to revise the meta model, and the above process is repeated. The revision of the meta model changes its composition according to the range of convergence and errors [12]. The history of the significance for the 16 iterations for each of the 3 constraint responses with respect to the design variables is displayed in Fig. 13. As seen the figure, the violation

Compared with base model, blue color components' thickness is decreased and red color components' thickness is increased like described in Fig. 14. Also, with the optimal design is satisfied for the 3 constraints (performance index), reducing the design objective function (BIW mass) as shown in Fig. 15.

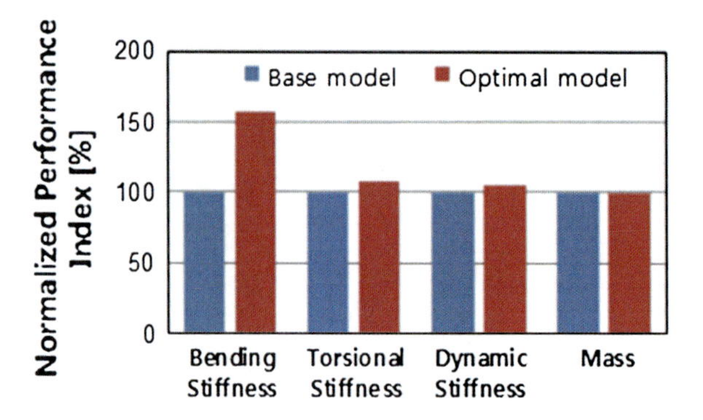

Fig. 15 Comparison of base model and optimal model performance index Graph

5 Conclusion

In the current research, a multi-disciplinary design approach for the reduction of vehicle body weight considering static stiffness (bending / torsional), dynamic stiffness (1st torsion mode) is suggested. Pursing this goal with an efficient application of design optimization, 13 design variables are carefully selected from about 200 design variables using the effect analysis and based on loadpath of load cases, and a meta model is constructed based on these selected design variables. The constructed meta model is optimized using the commercial software package EASY DESIGN and the results from the analysis are summarized as follows.

(1) Originally, the number of design variables was about 200. If all these variables are considered for the design optimization analysis regardless of the effectiveness that contributes the changes of the performance indices, it would take tremendous amount of calculation times. Therefore, in this research, in order to use the time efficiently, only 13 design variables were carefully selected based on the results from the effect analysis.
(2) The final solution was obtained through 16 iterations of numerical experiments running approximate design optimization calculations. The target function of body weight was found to be 0.02 % less than the base model.

Acknowledgments This research was supported by the International Collaborative Research Project between KSU and KMU and by the MKE (The Ministry of Knowledge Economy), Korea, under the CITRC (Convergence Information Technology Research Center) support program (NIPA-2012-H0401-12-2003) supervised by the NIPA(National IT Industry Promotion Agency).

References

1. Londhe, A., Kalani, D., Ali, A.: A systematic approach for weight reduction of BIW panels through optimization. SAE Technical Paper 2010-01-0389 (2010)
2. Mundoa, D., Hadjitb, R., Dondersb, S., Brughmansb, M., Masb, P., Desmetc, W.: Simplified modelling of joints and beam-like structures for BIW optimization in a concept phase of the vehicle design process. Finite Elem. Anal. Des. **45**(6–7), 456–462 (2009)
3. Jang, G.W., Choi, Y.M., Choi, G.J.: Discrete thickness optimization of an automobile body by using the continuous-variable-based method. J. Mech. Sci. Technol. **22**, 41–49 (2008)
4. Kim, B.J., Kim, M.S., Heo, S.J.: Aluminum space frame B.I.W. optimization considering multidisciplinary design constrains. Trans. KSAE **14**(1), 1–7 (2006)
5. Kodiyalam, S., Yang, R.J., Gu, L., Tho, C.H.: Multidisciplinary design optimization of a vehicle system in a scalable, high performance computing environment. Struct. Multi. Optim. **26**(3–4), 256–263 (2004)
6. Deb, A., Chou, C., Dutta, U., Gunti, S.: Practical versus RSM-based MDO in vehicle body design. SAE Technical Paper 2012–01-0098 (2012)
7. Gao, Y., Feng, H., Gao, D., Xu, R.: Multidisciplinary design optimization of a hatchback structure. SAE Technical Paper 2012–01-0780 (2012)
8. Yang, R., Chuang, C., Fu, Y.: An effective optimization strategy for structural weight reduction. SAE Technical Paper 2010–01-0647 (2010)
9. EasyDesign Training Guide: Institute of design optimization, Republic of Korea (2012)
10. Wang, J.G., Liu, G.R.: A point interpolation meshless method based on radial basis functions. Int. J. Numer. Methods Eng. **54**, 1623–1648 (2002)
11. Kim, M.S., Kim, C.W., Kim, J.H., Choi, J.H.: Efficient optimization method for noisy responses of mechanical systems. J. Mech. Eng. Sci. **222**(C), 2433–2439 (2008)
12. Kitayama, S., Arakawa, M., Yamazaki, K.: Sequential approximate optimization using radial basis function network for engineering optimization. Optim. Eng. **12**(4), 535–557 (2011)

Optimization Under Uncertainties

Rafael H. Lopez and André T. Beck

Abstract The goal of this chapter is to present the main approaches to optimization of engineering systems in the presence of uncertainties. That is, the chapter should serve as a guide to those entering in the exciting and challenging subject of optimization under uncertainties. First, the basic concepts of optimization and uncertainty quantification are presented separately. Then, the optimization under uncertainties techniques are described. We begin by giving an insight about the stochastic programming, also known as robust optimization in the engineering fields. Next, we detail how to deal with probabilistic constraints in optimization, the so called the reliability based design. Subsequently, we present the risk optimization approach, which includes the expected costs of failure in the objective function.

1 Introduction

Optimization has become a very important tool in several fields, for instance, investors seek to create portfolios that avoid excessive risk while achieving a high rate of return, manufacturers aim for maximum efficiency in the design and operation of their production processes, engineers adjust parameters to optimize the performance of their design and so on (Nocedal and Wright 2006). However, it is difficult to find examples of systems to be optimized that do not include some level of uncertainty

R. H. Lopez (✉)
Civil Engineering Department, Federal University of Santa Catarina, Rua João Pio Duarte Silva, s/n, Florianópolis, SC88040-900, Brazil
e-mail: rafael.holdorf@ufsc.br

A. T. Beck
Department of Structural Engineering, São Carlos School of Engineering, University of São Paulo, Av. Trabalhador São-carlense, 400, São Carlos, SP13566-590, Brazil
e-mail: atbeck@sc.usp.br

P. A. Muñoz-Rojas (ed.), *Optimization of Structures and Components*,
Advanced Structured Materials 43, DOI: 10.1007/978-3-319-00717-5_8,

about the values to assign to some of the parameters or about the actual design of some of the components of the system [77].

Nowadays it is widely acknowledged that deterministic optimization is not robust with respect to the uncertainties which affect engineering design. In deterministic optimization, potential failure modes of the designed engineering system are converted in design constraints, and uncertainty is addresses indirectly, by means of safety coefficients and conservative assumptions. This approach is inherited from design through design codes, which is essentially non-optimal (since approximations are always conservative). On the other hand, a deterministic optimal design has, naturally, more failure modes designed against the limit. Hence, deterministic optimal designs are potentially less safe than their non-optimal counterparts.

The acknowledgement that optimal design must be robust with respect to the uncertainties affecting its performance led to distinct philosophies of optimization under uncertainties, such as: expectation minimization, minimization of deviations from goals, minimization of maximum costs and inclusion of probabilistic constraints [65]. The main techniques used to take into account uncertainties in optimization are: stochastic programming [29], fuzzy variables [88] and reliability based design optimization [16].

Stochastic optimization addresses the minimization of some system responses such as the mean, the variance, and has been extensively employed in production planning, logistics, resource management, telecommunications, finances and risk management, among other areas.

In application to engineering problems, stochastic optimization has been called robust optimization [10, 64]. Typical objectives for robust optimization are maximization of mean performance and minimization of response variance. Generally, these are multi-objective problems [60, 66], whose solution involves weighted sums [17], compromise programming (Pareto fronts) [22] or preferential aggregation [48].

In stochastic or robust optimization, uncertainties are represented probabilistically, using random variables or stochastic processes. In fuzzy optimization, uncertainties are represented possibilistically, using intervals or fuzzy variables. Generally, the amount of available information guides the proper approach to be adopted. The probabilistic approach requires information on event probabilities, whereas lack of, vague or diffuse information leads to the possibilistic representation. The two types of uncertainty description can of course be combined in the same problem [8].

In Reliability-Based Design Optimization (RBDO), one looks for the minimization of some objective function involving material or manufacturing costs, or maximization of system performance, subject to restrictions on failure probabilities. The RBDO formulation explicitly addresses the uncertainties affecting system performance and, and ensures that a minimum, specified level of safety is maintained by the optimum design. Risk optimization increases the scope of the problem by including the expected costs of failure in the objective function. In this way, risk optimization allows one to address the compromising goals of economy and safety in the design of any engineering system.

The goal of this chapter is to present the main approaches to optimization of engineering systems in the presence of uncertainties. The chapter is organized as follows:

Sects. 2 and 3 present the fundamental formulations of optimization and uncertainty quantification. Sects. 4 addresses stochastic or robust optimization. RBDO is presented in Sects. 5, and risk optimization is presented in Sects. 6. Concluding remarks are presented in Sects. 7.

2 Optimization

The practical use of optimization processes begins with the definition of at least one objective, which is a measure of the performance of the system to be analyzed. It can be, for example, the minimization of a cost for obtaining a performance defined a priori. This objective depends on certain characteristics of the system, which are called design variables. The goal of the optimization is to find the values of the design variables that provide the best value of the objective. Sometimes these variables are limited to certain values what leads to a problem under constraints or a constrained optimization problem.

Thus, an optimization problem is defined by three elements:

- the objective: in general, a function associated with the parameters of the system under analysis that measures in some manner its performance. For example, a function $J : \Re^n \rightarrow \Re$ provides the value of the performance of the system associated with n parameters taking real values. An example is provided by the mass of a cylinder.
- the design variables, i.e., the parameters on which the designer can act to modify the characteristics that define the system to improve its performance. For example, a vector $\mathbf{d} = (d_1, \cdots, d_n) \in \Re^n$ can translate n adjustable parameters taking real values. The workspace associated is $V = \Re^n$. In the previous example, the design variables d_1 and d_2 can be respectively the radius and length of the cylinder.
- the design space, i.e., the constraints limiting the choices of the designer. In general, it is a non-empty subset S of V. In the example of the cylinder, a relationship between the design variables may be imposed, for example $0.5 \leq \frac{d_1}{d_2} \leq 2.0$.

An optimization problem may be posed in the following manner:

$$\mathbf{d}^* = \operatorname{argmin} \{J(\mathbf{d}, \mathbf{x}) : \mathbf{d} \in S\}, \tag{1}$$

in which:

- $J : \Re^n \times \Re^m \rightarrow \Re$ is the objective function to be minimized (e.g., cost, weight),
- $\mathbf{d} \in \Re^n$ are the design variables,
- $\mathbf{x} \in \Re^m$ are parameters of the system under analysis,
- $S = \{g_i(\mathbf{d}, \mathbf{x}) \leq 0, 1 \leq i \leq n_c, h_j(\mathbf{d}, \mathbf{x}) = 0, 1 \leq j \leq n_e\}$ is the design space or admissible set, $g_i : \Re^n \times \Re^m \rightarrow \Re$ and $h_j : \Re^n \times \Re^m \rightarrow \Re$ are inequality and equality constraints, respectively.

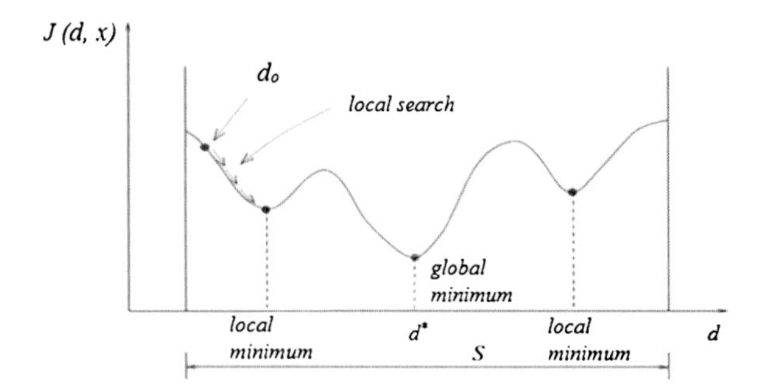

Fig. 1 Local and global minima

The problem consists in finding the global minimum of J, i.e., finding $\mathbf{d}^* \in S$ such that $J(\mathbf{d}^*, \mathbf{x}) \leq J(\mathbf{d}, \mathbf{x})\, \forall \mathbf{d} \in S$. Notice that if either J or S is not convex, there may exist local minima, i.e., $\mathbf{d}^* \in S$ such that $J(\mathbf{d}^*, \mathbf{x}) \leq J(\mathbf{d}, \mathbf{x})\, \forall \mathbf{d} \in S / \|\mathbf{d} - \mathbf{d}*\| \leq \varepsilon, \varepsilon > 0$.

In general situations, several global minima and many local minima may exist. Figure 1 shows an example of a single variable function with local minima and a unique global minimum. Among the minima, the one which has the smallest value of J is its global minimum.

Optimization algorithms are generally iterative. They start with an initial value $\mathbf{d}_0$ of the variable $\mathbf{d}$ and generate a sequence of points $\{\mathbf{d}_0, \mathbf{d}_1, \ldots, \mathbf{d}_{nit}\}$ which is supposed to converge towards the solution of the optimization problem. The strategy used to move from one point to the next is what distinguishes the different optimization algorithms. Most of the strategies proposed in the literature employ the values of the objective function, constraints and also first and second derivatives of these functions. Some algorithms accumulate the information collected during the iterations, while others only use the last computed information obtained at the current point. From this point of view, the algorithms can be classified according to the information they require. We have:

- Methods of order zero using only the value of the objective function and constraints. For example:
  - Dichotomy;
  - Nelder-Mead;
  - Standard genetic algorithms.
- First-order methods that use derivatives of objective functions and constraints. For example:
  - Gradient method;
  - Projected gradient method;
  - Method of penalties;
  - Augmented Lagrangian method.

- The second order methods that also use the information of the second derivative of these functions. For example: Newton's method.

Optimization algorithms can also be classified as:
local: we call "local" those methods which converge to a local minimum.
global: we call "global" those methods which are capable of converging to the global minimum.

Many local and global methods exist in the literature. Among the most popular local methods, we can mention: the descent methods, Newton methods, direct methods. Global methods often involve the use of probabilities (e.g. genetic algorithm, simulated annealing, random search) [39, 40]. We also consider hybrid algorithms consisting of a method combining local and global algorithms. These algorithms have been developed to reduce the cost of global search. We can cite, for example: "Global and Bounded Nealder-Mead" (Luersen and Le Riche 2003; [60] or the method of "random perturbation gradient" (Souza de Cursi 1994). For the interested reader, a general methodology for the hybridization of algorithms was presented by [70].

To compare optimization algorithms, we can consider three criteria (Nocedal and Wright 2006):

- Efficiency, which is identified by the number of calls to the objective function needed to achieve convergence;
- Robustness, which is defined by the ability of the algorithm to find the optimal point, regardless of the configuration of the problem, in particular, regardless of the starting point;
- Accuracy, which is the ability to find a specific solution, without being overly sensitive to data errors or rounding errors that occur when the algorithm is implemented on a computer.

These different elements are usually in conflict. For example, a method that converges quickly to a large unconstrained nonlinear problem may require too much storage memory. On the other hand, a robust method can also be slower. Compromise between the speed of convergence, storage requirements and robustness are the central questions of numerical optimization. It is important to point out that one of the difficulties inherent in optimization procedures lies in the fact that no universal algorithm is available. For example, for combinatorial problems, there is a theorem under the name of *No Free Lunch Theorem* (Wolpert and Macready 1997) which shows that there is no optimal algorithm for all problems. This theorem shows that any optimization method showing a good performance at one class of functions can have a bad performance for another class of problems. The responsibility to choose the algorithm suitable for a specific application is of the designer. The choice of algorithm is also an important step because it often determines the quality and speed of the problem resolution.

3 Uncertainities

Probability theory is the mathematical field that studies the phenomena characterized by randomness and uncertainty. The central objects of probability theory are random variables, stochastic processes, and events: they translate the abstract nondeterministic events or measured quantities that can sometimes evolve over time in an apparently random manner.

By the time of its development, the concept of probability was difficult to apply to complex systems, because the uncertainties were not easy to explain, formulate and manipulate digitally. The development of powerful computers has made possible the numerical solution of problems of large and/or complex systems subject to uncertainties.

Thus, the application of the theory of probability in all fields of engineering has grown in importance in recent decades. Researchers combined traditional design methods with uncertainty quantification methods. These new methods are able to provide robust designs even in the presence of uncertainties. Such stochastic analysis methods have been introduced in all areas of science and engineering (e.g., physics, meteorology, medicine, etc.).

Uncertainty can have several meanings: the probability of an event, the level of confidence, lack of knowledge, imprecision, variability, etc.. An accurate representation of uncertainties is crucial because different representations can lead to different interpretations of the system. The features and limitations of different representations have led to a classification of uncertainty into two categories: aleatory and epistemic.

3.1 Types of Uncertainty

The aleatory uncertainty (objective) is also called irreducible or inherent uncertainty. The epistemic uncertainty (subjective) is a reducible uncertainty caused by a lack of information and data. The birthday problem found in elementary probability books illustrates the difference between subjective and objective uncertainty: we present a person and we ask the question "What is the probability that this person was born on July 4?" An objective person answer is that the probability is 1/365. Another person, having more information about the nominated person (e.g., an acquaintance of the family) might have a different answer, e.g. 1/62 because he is sure that the birth is in July or August . The second person gives a higher probability (narrower limits) when compared to the first. The accuracy of the response depends on the degree of knowledge or the amount of information available. As subjective uncertainty is considered reducible according to the information available, based on past experience or expert judgment, it requires more attention and a more conservative estimation.

Four sources of uncertainty have been identified accordingly to [34]:

- Uncertainties on the model parameters: for example, the demand for a product is not known, the price of fuel fluctuates, the wind loading on a structure is not always the same. Uncertainties on the model parameters must be taken into account to improve the quality of its response.

- Uncertainties of models: we can still separate these uncertainties into two subgroups:
  - Uncertainties in the physical model,
  - Uncertainty about the finesse of representation.
- In practice, these two uncertainties are combined. A model can be based on different physical hypothesis (with or without fluid viscosity, linear or nonlinear behavior). In discrete models (finite element, boundary, finite differences), different models may correspond to different levels of representation of the geometry and / or mesh.
- Numerical uncertainties: the arithmetic calculators can cause uncertainties in the models.
- The uncertainty on the global solutions: in the absence of specific information on the function (e.g. Lipschitz regularity, number of local optima, unimodality problem), the global solutions found by an optimization method is uncertain, i.e., we cannot say that the minimum found is really an approximation of $\mathbf{d}^*$.

In this work, we restrict ourselves to uncertainties in model parameters. We present below characterization techniques for such uncertainties.

3.2 Uncertainty Representation

Two approaches are often used to represent uncertainties: the probabilistic approach and the possibilistic approach. The probabilistic approach is based on probability theory and involves the use of random variables, processes and fields to represent uncertainty [25]. In this approach, the probability density function (PDF) carries all the information relevant to the uncertainty. The PDF defines, for instance, the mean, variance, median.

In practice, the PDF of a given parameter in a real system is usually not completely available. However, if the interval of upper and lower bound of the random parameter is available, it can be used. The use of intervals to deal with uncertainties is called possibilistic approach and it describes better incomplete and imperfect data. Among the possibilistic approaches, we can mention: interval analysis [52], the fuzzy variables [88] and the evidence theory [68]. As an illustration, Fig. 2 shows the PDF and interval information.

3.3 Probabilistic Methods for Uncertainty Quantification

Methods of uncertainty quantification can be sorted into three broad categories:

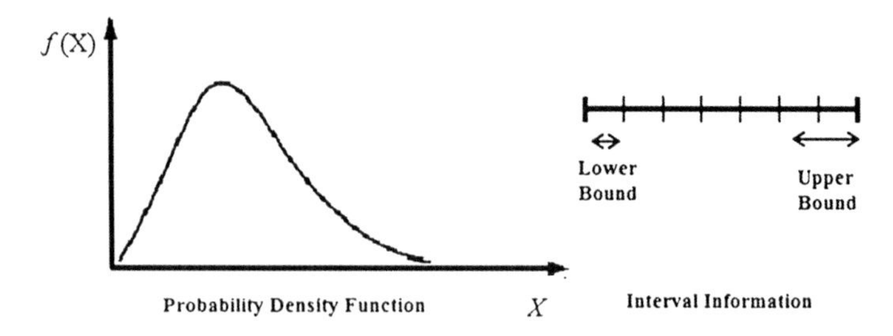

Fig. 2 Uncertainty representation. **a** probability density function, **b** interval information

- simulation methods,
- analytical methods,
- numerical integration.

The *simulation methods* using sampling and estimation are well known in the literature, the most widely used being the Direct Monte Carlo simulation (MCS) method [62]. The main drawback of MCS is that it requires a huge amount of calculations. Several improvements of the MCS have been developed to reduce the computational effort, such as the quasi-MCS [55], directional simulation [54] and importance sampling [24]. In practice, MCS is considered *the reference* response and is used to validate the results of other, approximate methods.

Analytical methods can be classified into three categories (Fig. 3), depending on the type of information they seek about the random variable:

- full characterization methods aim at obtaining the probability density function (PDF) of the response of the system [41, 78];
- response variability methods compute statistical moments of the response such as its mean and/or standard deviation [10];
- reliability methods investigate the probability of failure of the system [38].

The *numerical integration* methods employ multidimensional integrals to determine the probabilistic characteristics of the random response of the system. These integrals are evaluated numerically [67].

For a comparison of methods for quantifying uncertainty see ([35]). Now that we have presented the basics on optimization and uncertainty modeling, we pass to the subject of how to deal with uncertainties in an optimization problem.

3.4 Possibilistic Methods for Uncertainty Quantification

Among the several variants for possibilistic uncertainty representation methods, fuzzy variables and fuzzy logic is one of the most popular. Fuzzy logic was formalized by Lotfi Zadeh in 1965 and has been used in many fields such as automotive (ABS brakes), robotics (pattern recognition), the management of traffic (red lights),

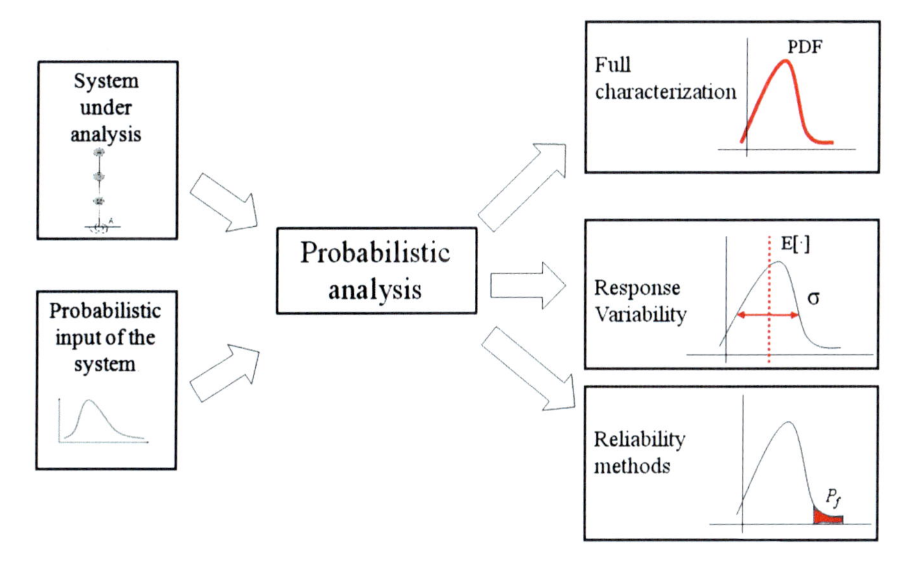

Fig. 3 Uncertainty quantification

air traffic control, environment (meteorology, climatology, seismology, life cycle analysis), medicine (diagnosis assistance), insurance (selection and risk prevention) and many others.

It is based on the mathematical theory of fuzzy sets. This theory, introduced by [86], is an extension of classical set theory for the inclusion of sets defined imprecisely. It is a formal and mathematical theory in the sense that Zadeh from the concept of membership function to model the definition of a subset of a given set, has developed a comprehensive model of formal definitions and properties. He also showed that the theory of fuzzy sets is effectively reduced to the theory of classical subsets where the membership functions are considered binary values ({0,1}).

Unlike Boolean logic, fuzzy logic allows a condition to be in a state other than true or false. There are degrees in checking a condition. For example, the membership function of a conventional A is defined by:

$$\vartheta_A(x) = \begin{cases} 1 & if\ x \in A \\ 0 & if\ x \notin A \end{cases} \tag{2}$$

This means that an element x is in A ($\vartheta_A(x)$= 1) or not ($\vartheta_A(x)$= 0). However, according to fuzzy logic, a fuzzy set A is defined by a membership function $\kappa_A(x)$: [0, 1]. Such a function associates to each element x in the fuzzy set A a degree of membership, it is the degree of compatibility between x e A. We then have:

$$\kappa_A(x) = \begin{cases} 1 & if\ x\ belongs\ to\ A; \\ 0 & if\ x\ does\ not\ belong\ to\ A; \\ [0,1] & if\ x\ belongs\ in\ part\ to\ A. \end{cases} \tag{3}$$

The main difference between stochastic programming and methods of fuzzy variables is how the system uncertainty is modeled. In stochastic programming, the uncertainties on the parameters are represented by random variables. The second approach considers the parameters as fuzzy variables and constraints as fuzzy sets. Both approaches can be used to solve the same problem. The selection is based on the knowledge available about the uncertainties of system under analysis. Thus, it is preferable to employ stochastic programming in situations where one knows the random parameters and that they can be represented by random variables. And, it is better to use fuzzy variables when uncertainty cannot be easily quantified [5]. The interested reader can consult [20] for more details on fuzzy logic and [51] for an overview of applications of engineering problems.

4 Stochastic Programming or Robust Design

Stochastic programming can be seen as an extension of linear programming and non-linear decision models where the coefficients are not known, i.e. they are represented probabilistically. Stochastic programming generally deals with issues that require a decision without further observations of the distributions of random parameters [77]. For example, a well-known problem in stochastic programming is the stochastic optimization with recourse, where decisions must be taken at the first stage (production order) to maximize gains and minimize the expected cost of the second stage (costs of rearing and overproduction are not known at the first stage, for example). A stochastic program can be given by:

$$\mathbf{d}^* = \operatorname{argmin}\left\{J(\mathbf{d}, \mathbf{X}) = c^T\mathbf{d} + E\left[Q(\mathbf{d}, \mathbf{X})\right] : \mathbf{d} \in S \subset \Re^n\right\}, \tag{4}$$

where $c \in \Re^n$ is the vector associated with the cost, $\mathbf{X}$ is a random vector whose support is $\Sigma \subset \Re^m$, and $Q(\mathbf{d}, \mathbf{X})$ is the cost associated to the second stage, which is random. Note that the objective function to be minimized is a function of the random vector $\mathbf{X}$, so it becomes itself a random variable. The most common techniques used to solve stochastic programming problems are stochastic approximation [30, 61], methods based on simulation [69] and decomposition [31]. References [11, 29] can be consulted for more information on the formulation and resolution of such type of problem. Stochastic programming has been widely applied to production planning, telecommunication and finances, to name just a few areas.

In the field of engineering, stochastic programming is better known as *robust optimization* or *robust design*. In this area, a key aspect is the definition of robustness, which may be defined as an optimal design that is not very sensitive to variability on the system parameters. It is generally expressed in mathematical form by one or a combination of statistical properties of the function to be minimized.

A very important aspect in robust optimization is the mathematical definition of robustness, i.e. the choice of the *robustness measure*. In addition, we illustrate how the objective function and the constraints of the problem are changed. Several measures of robustness have been proposed in the literature [13, 19, 60], for example :

- Simultaneous minimization of the mean and variance of J:

$$J_{sto}(\mathbf{d}, \mathbf{X}) = CCS\{E[J(\mathbf{d}, \mathbf{X})], \text{var}[J(\mathbf{d}, \mathbf{X})]\} \tag{5}$$

- Simultaneous minimization of the mean and variance of the difference between J and a target performance $\hat{J}$

$$J_{sto}(\mathbf{d}, \mathbf{X}) = CCS\left\{E\left[J(\mathbf{d}, \mathbf{X}) - \hat{J}\right], \text{var}\left[J(\mathbf{d}, \mathbf{X}) - \hat{J}\right]\right\} \tag{6}$$

- Minimization of the ith percentile of J

$$J_{sto}(\mathbf{d}, \mathbf{X}) = Cen_i[J(\mathbf{d}, \mathbf{X})], \tag{7}$$

where J_{sto} is the objective function of the robust optimization problem, CCS is a chosen function and Cen_i is ith percentile of J_{sto}.

As pointed out above, several objectives may be required to be fullfilled simultaneously in robust optimization (typically, minimizing the mean and variance of a variable, see Eq. (5)). In this case, a multi-objective optimization problem (MOP) can be formulated [4, 66]. A MOP may be stated as:

$$\mathbf{d}^* = \text{argmin}\left\{\mathbf{J}_{sto}(\mathbf{d}, \mathbf{X}) = \left(J_1(\mathbf{d}, \mathbf{X}), \ldots, J_{nf}(\mathbf{d}, \mathbf{X})\right) : \mathbf{d} \in S \subset \Re^n\right\}, \tag{8}$$

where $\mathbf{J}_{sto} : S \rightarrow \Im \subset \Re^{nf}$ consists of m real-valued objective functions and $\Im$ is called the objective space. Usually, the objectives contradict each other and, consequently, no point in S minimizes all of them simultaneously. The best tradeoffs among the objectives are defined in terms of Pareto optimality, and the main goal of the MOP is to construct the Pareto front [3].

The Pareto front may be approximated as the solution of a series of scalar optimization subproblems in which the objective is an aggregation of all the J_i's. Several methods are found in the literature for constructing aggregation functions, such as the weighted sum, Tchebycheff inequality, the normal boundary intersection, the normal constraint method, the Physical Programming method, the epsilon constraints and Directed Search Domain [15, 18, 47–49, 87]. The development of MOP methods has sought the generation of an even spread of Pareto optimal points as well as the treatment of the non-convexity on both the Pareto front and the scalar optimization subproblems.

A second aspect concerns the constraints in the robust optimization problem (i.e., the admissible set). When the solution is close to the boundaries of the admissible set, it is essential to ensure that the fluctuations of random parameters do not lead to non-admissible points which may correspond to the failure of the system. In this case, the design is considered robust if it remains eligible with a probability chosen by the designer even under the effect of the uncertainties of the system [56]. Approaches to address this situation can be divided into two groups: methods that require a probabilistic analysis (MCS, Latin Hypercube methods, First and Second Order Reliability Methods - FORM and SORM, respectively) or statistical methods which do not require such an analysis (e.g., worst case scenario) [22]. The methods of the first group are addressed in the RBDO section. The methods of the second group will not be addressed herein. The interested reader is referred to [22] for a comprehensive overview of such methods.

In what follows, we describe a classical approach used in robust optimization: the Taguchi method [73].

4.1 The Taguchi Method

The first attempts to account for uncertainties in the design, as part of the quality engineering, are closely related to Taguchi. In the fifties and early sixties, Taguchi developed the basis of the robust design to produce high quality products. In 1980, he applied his methods in the field of telecommunications in the American industry. Since then, the Taguchi method has been successfully applied to various industries such as electronics, automotive and telecommunications [57, 71, 72].

The main difference between the Taguchi method and the deterministic optimization is the performance measure due to random parameters and variations. The Taguchi approach considers two types of parameters in the objective function: control parameters or design variables $\mathbf{d}$, which must be chosen in order to find the optimal point, and noise factors $\mathbf{X}$, such as environmental conditions (temperature , pressure, etc.) and production tolerances (for example, the change in weight and length, etc.) which are hardly controlled by the designer.

The aim of the Taguchi method is to find a system configuration that is insensitive to the effects of variability without eliminating them. Figure 4 illustrates the different types of variations in performance. The large circles represent the performance targets and the dots indicate the distribution of the response. The robustness measure in this case may be the simultaneous minimization of the mean and variance of the difference between system performance and the target performance. In other words, the goal of robust design here is to make the system performance close to the performance target, with slight variations, as shown in Fig. 4d.

Taguchi developed his design methodology in three steps [73]:

- Modeling: determining the basic performance parameters of the product and its general structure (design variables $\mathbf{d}$, random parameters $\mathbf{X}$, performance function $\mathbf{y}$).
- Determination of parameters: optimize design variables in order to match quality requirements. That is to say, to achieve the target design.
- Taking into account the tolerance: fine-tune the design variables obtained in the second step.

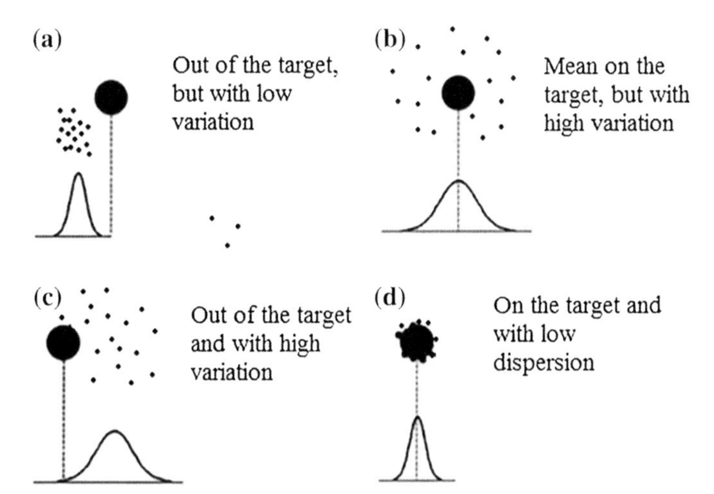

Fig. 4 Different types of the variability ofthe the system performance. **a** Out of the target but with low variation, **b** Mean on the target, but with high variation, **c** Out of the target and with high variation, **d** On the target and with low dispersion

Depending on the purpose of the design, Taguchi proposed the signal to noise ratios. Consider $y_i = (\mathbf{d}, \mathbf{X}_i)$ the performance value of a single sample of the random vector. Equation (2) measures the variance of y with respect to the target performance $\hat{y}$.

$$MSD_1 = \frac{1}{n_s}\sum_{i=1}^{n_s}\left(y\left(\mathbf{d}, \mathbf{X}_i\right) - \hat{y}\right)^2, \tag{9}$$

If the target performance is close to zero, Eq. (9) results in:

$$MSD_2 = \frac{1}{n_s}\sum_{i=1}^{n_s}\left(y\left(\mathbf{d}, \mathbf{X}_i\right)\right)^2, \tag{10}$$

On the other hand, Taguchi proposed:

$$MSD_2 = \frac{1}{n_s}\sum_{i=1}^{n_s}\left(y\left(\mathbf{d}, \mathbf{X}_i\right)\right)^{-2}, \tag{11}$$

Using the MSD functions, Taguchi defined the signal to noise ratio:

$$SNR = -10\log_{10}(MSD) \tag{12}$$

which becomes the objective function to be maximized. Taguchi approach used the design of experiments [63] to perform this optimization.

Although the contributions of Taguchi to robust design are almost unanimously considered of fundamental importance, they are affected by limitations and inefficiencies intrinsically linked to its approach. For example, the optimization method used does not always find the optimal solution for the considered system [12] and the method is not necessarily an accurate solution for strongly nonlinear problems [75]. Other controversial aspects of the Taguchi method are summarized in [53].

5 Reliability Based Design Optimization

Deterministic Design Optimization (DDO) grossly neglects the effects of parameter uncertainty and failure consequences, but it allows one to find the shape or configuration of a structure that is optimum in terms of mechanics. As a general rule, the result of DDO is a structure with more failure modes designed against the limit: hence, the optimum structure compromises safety, in comparison to the original (non-optimal) structure. RBDO has emerged as an alternative to properly model the safety-under-uncertainty aspect of the optimization problem. With RBDO, one can ensure that a minimum (and measurable) level of safety, chosen *a priori* by the designer, is maintained by the optimum structure.

The RBDO problem may be stated as:

$$\left| \begin{array}{ll} \text{Minimize:} & J(\mathbf{d}, \mathbf{X}) \\ \text{subject to:} & P_{f_i} = P\left(G_i\left(d, X\right) < 0\right) \leq P_{f_i}^{allowable} \; i = 1 \ldots n_c \end{array} \right. \tag{13}$$

where $\mathbf{d} \in \Re^n$ is the design vector (e.g. structural configuration and dimensions), $\mathbf{X} \in \Re^m$ contains all the random variables of the system under analysis (e.g. random loads, uncertain structural parameters), J is the objective function to be minimized (e.g. the structural weight, volume or manufacturing cost), P_{f_i} is the probability of failure of ith constraint (G_i), and $P_{f_i}^{allowable}$ is the allowable (maximum) failure probability for the ith constraint.

The failure probability P_{f_i} for each constraint may be obtained by evaluating the integral in Eq. (2), which is the fundamental expression of the structural reliability problem:

$$P_{f_i} = \int_{G_i(\mathbf{d},\mathbf{X})<0} f_{\mathbf{X}}(\mathbf{x})\, d\mathbf{x} \tag{14}$$

where $f_{\mathbf{X}}(\mathbf{x})$ is the joint probability density function (PDF) of random vector $\mathbf{X}$. In practice, it is impossible to obtain the joint PDF because of scarcity of joint observations for a large number of random variables. At best, what is known are the marginal probability distributions of each random variable and possibly correlations between pairs of random variables. Another difficulty in solving Eq. (14) is the fact that the limit state equations, G_i, are sometimes given in implicit form, as the response of finite element models (Beck and da Rosa 2006). Such difficulties have motivated the development of various approximate reliability methods.

Among these methods, the most widely applied approaches are the First and Second Order Reliability Methods (FORM and SORM, respectively). FORM and SORM are said to be transformation methods, because the integral in Eq. (14) is not solved in the original space ($\boldsymbol{X}$), but is mapped to the Standard Gaussian space. The main advantage of FORM-based approaches is their computational cost, which is a fraction of the cost of crude MCS, for instance.

Hence, FORM based approaches have been widely employed to evaluate Eq. (4) in RBDO problems. As detailed in following sections, FORM is an optimization problem itself and consequently, the RBDO using FORM is a double-loop strategy:

- the inner loop is the reliability analysis,
- the outer loop the structural optimization,

i.e. the two optimizations are coupled. Such coupling of optimization loops: structural optimization and reliability assessment–leads to very high computational costs. To reduce the computational burden of RBDO, several authors decoupled the structural optimization and the reliability analysis. Techniques for de-coupling the optimization loops may be divided in two groups: (1) serial single loop methods and, (2) unilevel methods.

The next sections focus on RBDO methods based on FORM. First, the two main FORM based approaches, Reliability Index Approach (RIA) and Performance Measure Approach (PMA), are described and compared Sect. 5.1. Then, the RBDO problem based on these approaches is detailed in Sect. 5.2. A review on decoupling methodologies is presented in Sect. 5.3.

5.1 Coupled FORM-Based Approaches

This section briefly details the approximation of Eq. (4) using RIA and PMA, then presents a review of the comparison between these two approaches. The interested reader is referred to [45, 46] and [27] for more details on the RIA, and to [76] for a full description of the PMA.

5.1.1 The Iso-Probabilistic Transformation

In order to approximate the integral in Eq. (4), it is usual to introduce a vector of normalized and statistically independent random variables $\mathbf{U} \in \Re^m$ and a transformation T(Fig. 5), so that $\mathbf{U} = T(\mathbf{X})$. The most common transformations are the Rosenblatt and the Nataf ones [38, 46]. The mapping T transforms every realization $\mathbf{x}$ of $\mathbf{X}$ in the physical space into a realization $\mathbf{u}$ in the normalized space. Note that it also holds for the constraints:

$$G_i(\mathbf{d}, \mathbf{X}) = G_i\left(\mathbf{d}, T^{-1}(\mathbf{d}, \mathbf{U})\right) = g_i(\mathbf{d}, \mathbf{U}), \tag{15}$$

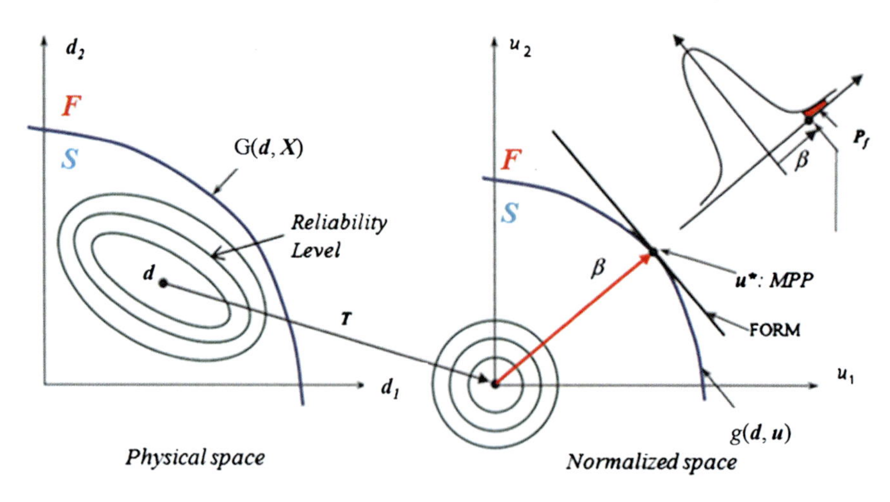

Fig. 5 Transformation T and the first order reliability method (FORM). **a** Physical space, **b** Normalized space

where g_i is the ith constraint in the normalized space.

The main advantage in the use of this transformation is that the probability distribution on the resulting space depends only on its norm. This fact is illustrated in Fig. 5 by the circular reliability levels of the normalized space. It must be highlighted that this transformation is the first approximation proposed to solve Eq. (4): it only is exact when **X** is comprised by independent Gaussian random variables. However, even with this simplification, it is still not an easy task to evaluate Eq. 4. The FORM approximation is used to simplify this evaluation.

5.1.2 Reliability Index Approach and Performance Measure Approach

The main idea of the FORM is simple: it consists in replacing the limit state function G_i by a tangent hyper-plane at the most probable point of failure (MPP). Figures 5 and 6 illustrate the approximation made by this hyper-plane. The FORM approximates the probability of failure and the allowable failure probability for the ith constraint by:

$$P_{f_i} \approx \Phi(-\beta_i) \quad \text{and} \quad P_{f_i}^{\text{allowable}} \approx \Phi\left(-\beta_i^{\text{target}}\right), \tag{16}$$

where Φ is the standard Gaussian cumulative distribution function (CDF) and β_i^{target} is the target reliability index for the ith constraint. This is the second approximation proposed to solve Eq. 4 and it only provides an exact result when the constraint is linear.

Now, recall that in order to evaluate β_i, one first needs to obtain the MPP: the point in the failure domain closest to the origin of the normalized space. The difference

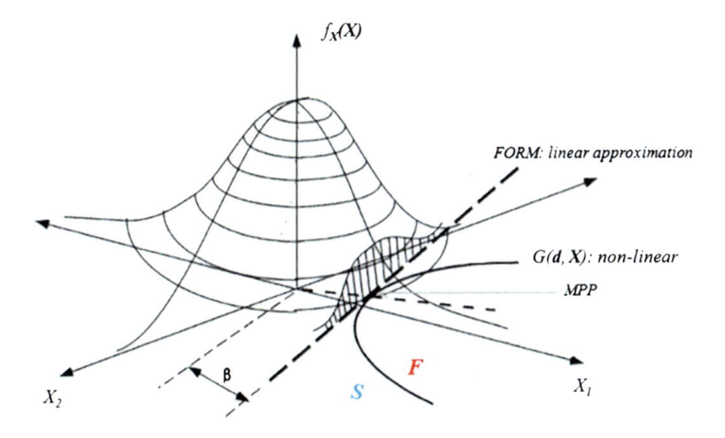

Fig. 6 Illustration of the FORM approximation

between the RIA and the PMA is the manner in which the MPP is calculated. The following optimization problems detail this difference:

RIA	**PMA**
for a given design d	for a given design d
find $\mathbf{u}^*_{RIA^i}$ which	find $\mathbf{u}^*_{PMA^i}$ which
minimizes : $\|\mathbf{u}\| = \beta_i$	minimizes : $g_i(\mathbf{d}, \mathbf{u})$
subject to : $g_i(\mathbf{d}, \mathbf{u}) = 0$	subject to : $\|\mathbf{u}\| = \beta_i^{\text{target}}$

The optimal solution $\mathbf{u}^*_{RIA^i}$ of the RIA yields the reliability index $\beta_i = \left\|\mathbf{u}^*_{RIA_i}\right\|$ of the ith constraint on the current design $\mathbf{x}$. On the other hand, the optimal solution $\mathbf{u}^*_{PMA^i}$ of the PMA is the minimum performance target point (MPTP) on the target reliability sphere (defined by $\|\mathbf{u}\| = \beta_i^{\text{target}}$) and it provides the so-called performance measure $p_{m_i} = g_i\left(\mathbf{d}, \mathbf{u}^*_{PMA_i}\right)$ of the ith constraint on the current design $\mathbf{d}$. The performance measure p_{m_i} is related to the reliability index β_i by the following relation:

$$p_{m_i} = F_{G_i}^{-1}(\Phi(-\beta_i)) \tag{17}$$

where F_{G_i} is CDF of the ith constraint. It is important to note that $\mathbf{u}^*_{RIA^i}$ and $\mathbf{u}^*_{PMA^i}$ will be equal only when the reliability constraint is active, e.g., at the final design of a RBDO problem. At any other point, $\mathbf{u}^*_{PMA^i}$ only represents the point of minimal performance on the target reliability sphere.

5.1.3 RIA Versus PMA

In the paper that introduced the PMA, [76] showed that PMA is inherently robust and yields a higher overall RBDO rate of convergence when compared to the RIA. [83], although reaching the same conclusions, showed that the PMA is far more effective when the probabilistic constraint is either very feasible or very infeasible. In a different paper, Youn and Choi (2004a) concluded that the PMA is quite attractive when compared to other probabilistic approaches in RBDO, such as the RIA and the approximate moment approach [36].

The first main difference between the RIA and the PMA is the type of optimization problem which is solved in each case. It is easier to minimize a complicated function subject to a simple constraint (PMA) than to minimize a simple function subject to a complicated constraint (RIA). Different from the RIA, in the PMA only the direction vector needs to be determined by taking advantage of the spherical equality constraint $\|\mathbf{u}\| = \beta_i^{\text{target}}$ to find the MPP$\mathbf{u}^*_{PMAi}$.

Still regarding the type of optimization problem, the conceptual iteration history during the search facilitates the PMA. Usually, the RIA search requires several iterations to reach the failure surface given by $g_i(\mathbf{d}, \mathbf{u}) = 0$, while the PMA search immediately lies on the $\|\mathbf{u}\| = \beta_i^{\text{target}}$ sphere; in other words, the number of iterations of RIA increases with the reliability index while the PMA search is independent of the target performance [36]. A second consequence is that, in the case of non-activated constraints, the PMA becomes even more effective.

Regarding non-linearities in the RBDO problem (i.e. use of non-normal random variables), [84] showed that PMA is more stable, efficient and has a lower dependence on the distribution of the random variables, since it introduces small non-linearities in the space-transformations. PMA can thus handle a variety of distributions without significantly increasing the number of function evaluations. Furthermore, RIA diverged when uniform or Gumble random variables were employed. The former divergence was due to the fundamental nature of the uniform distribution and the latter was due to numerical difficulties when dealing either with a nonlinear failure surface or with a failure surface away from the design point.

5.2 *Classical Coupled Approaches*

As the reliability analysis is an optimisation procedure by itself, RBDO, in its classical version, is a double-loop strategy: the inner loop is the reliability analysis and the outer loop is the structural optimisation. Thus, the two optimisation loops are coupled:

for $k = 1, 2, \ldots$

(a) structural optimization:

$$\begin{aligned}
\text{minimize:} \quad & J\left(\mathbf{d}^{(k)}\right) \\
\text{subject to:} \quad & \beta_i\left(\mathbf{d}^{(k-1)}\right) + \left(\nabla_{\mathbf{d}}\beta_i\left(\mathbf{d}^{(k-1)}\right)\right)^T\left(\mathbf{d}^{(k)} - \mathbf{d}^{(k-1)}\right) \le 0(\text{RIA}), \\
& p_{mi}\left(\mathbf{d}^{(k-1)}\right) + \left(\nabla_{\mathbf{d}} p_{mi}\left(\mathbf{d}^{(k-1)}\right)\right)^T\left(\mathbf{d}^{(k)} - \mathbf{d}^{(k-1)}\right) \le 0(\text{PMA}), \\
& i = 1 \ldots n_{c,} \\
& \mathbf{d}_l \le \mathbf{d}^{(k)} \le \mathbf{d}_u
\end{aligned}$$

where, at each step k, given current design $\mathbf{d}^{(k-1)}$, the reliability routine is called:

(b) reliability analysis:

	RIA	**PMA**
	find $\mathbf{u}^{*(k-1)}_{RIA_i}$ which	find $\mathbf{u}^{*(k-1)}_{PMA_i}$ which
minimizes:	$\lVert\mathbf{u}\rVert = \beta_i$	$g_i\left(\mathbf{d}^{(k-1)}, \mathbf{u}\right)$
subject to:	$g_i\left(\mathbf{d}^{(k-1)}, \mathbf{u}\right) = 0$	$\lVert\mathbf{u}\rVert = \beta_i^{\text{target}}$

At the end of each reliability analysis, a sensitivity analysis of the design variables with respect to the reliability index is pursued to obtain $\nabla_{\mathbf{d}}\beta_i\left(\mathbf{d}^{(k-1)}\right)$or $\nabla_{\mathbf{d}} p_{m_i}\left(\mathbf{d}^{(k-1)}\right)$. This procedure is repeated until some convergence criterion is achieved and, of course, it leads to very high computational costs. A review of techniques developed to de-couple the RBDO problem, in order to reduce the computational burden, is presented in the next section.

5.3 Decoupling Strategies

De-coupling the two optimization problems means not to have to call the reliability analysis routine at every step k of the structural optimization. In the sequel, the serial single loop and unilevel de-coupling methods are reviewed.

5.3.1 Serial Single Loop Methods

The basic idea of the serial single loop methods is to decouple the structural optimization (outer loop) and the reliability optimization (inner loop). Each method of single-loop decoupling employs a specific strategy to decouple the loops and then solves them sequentially until some convergence criterion is achieved. Among these methods, the following may be cited: Traditional Approximation Method (TAM), Single Loop Single Variable (SLSV), Sequential Optimisation and Reliability Assessment (SORA) and Safety Factor Approach (SFA) (Lopez and Beck 2012).

Yang and Gu [80] compared these four single-loop decoupling RBDO methods. Four different examples were solved including a vehicle side impact and a multidisciplinary optimization problem. According to their results, SLSV was the most effective method, converging nicely and requiring the fewest number of function evaluations. The other methods also showed promising results when compared to the classical approach. In a second paper, the authors [79] investigated an engineering problem with a large number of constraints (144) and with many local minima. Results showed that the number of function evaluations depends on the RBDO method, optimization algorithm and implementation. Furthermore, algorithms with good active-constraint handling tended to perform better (e.g. SORA/SFA). Moreover, SORA/SFA and TAM have advantages over the other methods, as the target reliability is obtained at the end. Regarding the local minima, different methods and different starting points yielded different final results, since only local optimizers were used by the authors.

5.3.2 Unilevel Methods

The central idea of unilevel methods is to replace the reliability analysis by some optimality criteria on the optimum (i.e. imposing it as a constraint in the outer loop). Thus, there is a concurrent convergence of the design optimization and reliability calculation or, in other words, they are sought simultaneously and independently.

Kuschel and Rackwitz [33] formulated an unilevel method based on replacing the inner loop of the classical approach (FORM analysis using RIA) by the first order Karush-Kuhn-Tucker (KKT) optimality conditions of the first-order reliability problem. In other words, the KKT optimality conditions of the RIA search are imposed as constraints in the outer loop of the RBDO. As already commented, the RIA may be ill-conditioned when the probability of failure given by a constraint is zero and it is not computationally efficient when the reliability index is large. With this in mind, [1] proposed a unilevel RBDO method which introduced the first order KKT necessary optimality conditions of PMA as constraints in the outer loop, eliminating the costly reliability analysis (inner loop) of RBDO.

Cheng et al. [14] proposed a unilevel strategy based on the Sequential Approximate Programming (SAP) concept which was successfully applied in structural optimization. In the SAP approach, the original optimization problem is decomposed into a sequence of sub-optimization problems. Each sub-optimization problem consists of an approximate objective function subjected to a set of approximate constraint functions. A SAP strategy for RBDO using the PMA to approximate the reliability constraints was also developed [81, 82].

Studies comparing the different unilevel methods have not been found in the literature. [82] compared the SAP based on RIA and SAP based on PMA with SORA and SLSV methods. Several examples, including the 144 constraint problem, were solved. Based on the results, the authors concluded that SAP based on PMA achieved better results than the other methods. This result does not imply that SAP-PMA is the most effective method in all cases; but it is, at least, one of the most powerful algorithms in RBDO.

6 Risk Optimization

In comparison to the RBDO approach, Risk Optimization (RO) increases the scope of the problem, by including expected costs of failure in the objective function. Expected costs of failure are obtained by multiplying failure costs by failure probabilities (Eq. 4). Hence, in terms of formulation, the fundamental difference between RBDO and RO is that in RBDO the failure probability is a constraint, whereas in RO the failure probability is part of the objective function.

The RO problem may be stated as:

$$Minimize : J(\mathbf{d}, \mathbf{X}) + \sum_{i=1}^{n_c} P_{f_i} J_{f_i} \tag{18}$$

where $\mathbf{d} \in \Re^n$ is the design vector (e.g. structural configuration and dimensions), $\mathbf{X} \in \Re^m$ contains all the random variables of the system under analysis (e.g. random loads, uncertain structural parameters), J is the (original, RBDO) objective function to be minimized (e.g. the structural weight, volume or manufacturing cost), P_{f_i} is the probability of failure for the ith constraint (G_i) and J_{f_i} is the cost of exceeding the ith constraint (cost of failure).

The risk optimization problem may be posed as:

$$\mathbf{d}^* = \text{argmin} \left\{ J(\mathbf{d}, \mathbf{x}) + \sum_{i=1}^{n_c} P_{f_i}(\mathbf{d}, \mathbf{x}) J_{f_i} : \mathbf{d} \in S \right\}, \tag{19}$$

in which S are simple restrictions on the allowable values for the design variables.

In engineering design, economy and safety are generally considered to be competing goals. Indeed, increasing safety implies greater immediate costs, and reducing costs may compromise safety. Hence, designing structural systems would involve a tradeoff between safety and economy. In common engineering practice, this tradeoff is addressed subjectively. When using structural design codes, the tradeoff has already been decided by a code committee, which specifies safety coefficients to be used in design, and basic safety measures to be adopted in construction and operation. In deterministic optimization, this tradeoff is completely neglected, because failure probabilities are not quantified. In classical Reliability-Based Design Optimization (Sect. 5) the tradeoff between safety and economy is also not addressed, because failure probabilities are constraints and not design variables. Robust design optimization (Sect. X.4) searches for designs which are less sensitive to the existing uncertainties, but safety-economy tradeoffs are also not addressed.

However, when expected costs of failure are included in the design equation, one realizes that economy and safety are not competing goals. Safety is just another design variable which directly affects expected costs of failure. Since failure probabilities and consequences of failure are directly affected by structural design, optimum (minimum cost) design can only be achieved by quantifying uncertainties, probabilities of failure and costs of failure. In other words, optimum (minimum cost) design can only be achieved by quantifying expected costs of failure, and by treating safety as a design variable. This is called structural risk optimization herein and in a few other references [7–9]. Risk optimization is a tool for decision making in the presence of uncertainty.

It has been observed that including failure probabilities in the objective function (Eq. 3) leads to problems with several local minima [7, 26]. Hence, global optimization algorithms have to be employed in the solution of RO problems.

The different approaches to optimization under uncertainties have been addresses in isolation in this paper. However, these approaches may be combined in actual applications. For instance, in [8] the robust optimization approach is combined with the risk optimization approach, in order to obtain robust minimum cost designs. These designs are robust in the sense that they are made less sensitive to uncertainties of the phenomenological type. In the article, probabilistic and possibilistic (fuzzy variable) types of uncertainty descriptions are also combined.

7 Concluding Remarks

This chapter presented a conspectus of philosophies and techniques for optimization in the presence of uncertainties. As uncertainties are ubiquitous in engineering endeavors, their explicit consideration leads to more robust, safer and more economical optimum designs. The chapter covered stochastic or robust optimization, reliability-based design optimization and risk optimization. The fundamental differences in the modeling assumptions and in results that can be achieved with each formulation have been highlighted. The chapter should serve as a guide to those entering in the exciting and challenging subject of optimization under uncertainties.

References

1. Agarwal, H., Mozumder, C.K., Renaud, J.E., Watson, L.T.: An inverse-measure-based unilevel architecture for reliability-based design. Structural and Multidisciplinary Optimization **33**, 217–227 (2007)
2. Aoues, Y., Chateauneuf, A.: Benchmark study of numerical methods for reliability-based design optimization. Structural and Multidisciplinary Optimization **41**, 277–294 (2010)
3. Arora, J.S.: Introduction to optimum design. Elsevier, London (2004)
4. Athan, T.W., Papalambros, P.Y.: A note on weighted criteria methods for compromise solutions in multi-objective optimization. Engineering Optimization **27**, 155–176 (1996)
5. Bastin, F.: Trust-region algorithms for nonlinear stochastic programming and mixed logit models. Thesis (PhD), Facultes Universitaires Notre-Dame de la Paix Namur (2004)
6. Beck, A.T., da Rosa, E.: Structural Reliability Analysis Using Deterministic Finite Element Programs. Latin American Journal of Solids and Structures **3**, 197–222 (2006)
7. Beck, A.T., Gomes, W.J.S.: A Comparison of Deterministic, Reliability-Based and Risk-Based Design Optimization. Probabilistic Engineering Mechanics **28**, 18–29 (2012)

8. Beck, A.T., Gomes, W.J.S., Bazán, F.A.V.: On the Robustness of Structural Risk Optimization with Respect to Epistemic Uncertainties. International Journal for Uncertainty Quantification **2**, 1–20 (2012)
9. Beck, A.T., Verzenhassi, C.C.: Risk Optimization of a Steel Frame Communications Tower Subject to Tornado Winds. Latin American Journal of Solids and Structures **5**, 187–203 (2008)
10. Beyer, H.G., Sendhoff, B.: Robust optimization - a comprehensive review. Computational Methods in Applied Mechanics and Engineering **196**, 3190–3218 (2006)
11. Birge, J.R., Louveaux, F.V.: Introduction to Stochastic Programming. Springer, New York (1997)
12. Box G, Fung C (1986) Studies in quality improvement: minimizing transmitted variation by parameter design. In: Report No. 8, Center for Quality Improvement, University of Wisconsin.
13. Capiez-Lernout, E., Soize, C.: Design optimization with an uncertain vibroacustic model. Journal of Vibration and Acoustic **130**(021001–1), 021001–8 (2008)
14. Cheng, G.D., Xu, L., Jiang, L.: Sequential approximate programming strategy for reliability-based optimization. Computers and Structures **84**, 1353–67 (2006)
15. Chinchuluun, A., Pardalos, P.M.: A survey of recent developments in multiobjective optimization. Annals of Operations Research **159**, 29–50 (2007)
16. Choi, S.K., Grandhi, R.V., Canfield, R.A.: Reliability-based structural design. Springer-Verlag, London (2007)
17. Das, I., Dennis, J.E.: A closer look at drawbacks of minimizing weighted sums of objectives for Pareto set generation in multi-criteria optimization problems. Structural Optimization **14**, 63–69 (1997)
18. Das, I., Dennis, J.E.: Normal-boundary intersection: a new method for generating Pareto optimal point in nonlinear multicriteria optimization problems. SIAM Journal on Optimization **8**, 631–657 (1998)
19. Doltsinis, I., Kang, Z.: Robust design of structures using optimization methods. Computational Methods in Applied Mechanics and Engineering **193**, 2221–2237 (2004)
20. Du Bois, D.: Fuzzy sets and systems: theory and applications. Academic Press, New York (1980)
21. Du, X., Chen, W.: Sequential Optimization and Reliability Assessment method for Efficient Probabilistic Design. ASME J Mech Des **126**, 225–233 (2004)
22. Du, X., Chen, W.: Towards a better understanding of modeling feasibility robustness in engineering design. Journal of Mechanical Design **122**, 385–394 (2000)
23. Du, X.: Saddlepoint Approximation for Sequential Optimization and Reliability Analysis. ASME J Mech Des **130**(011011), 1–11 (2008)
24. Engelund, S., Rackwitz, R.: A benchmark study on importance sampling techniques in structural reliability. Struct Safety **12**, 255–76 (1993)
25. Feller, W.: An Introduction to Probability Theory and its Applications. Wiley, New York (1968)
26. Gomes WJS, Beck AT (2013) Global structural optimization considering expected consequences of failure and using ANN surrogates. to appear in Computers & Structures (accepted paper), dx.doi.org/10.1016/j.compstruc.2012.10.013.
27. Haldar, A., Mahadevan, S.: Reliability Assessment Using Stochastic Finite Element Analysis. John Wiley & Sons, New York (2000)
28. Hasofer AM, Lind NC (1974) An exact and invariant first order reliability format. Journal of the Engineering Mechanics Division. 100(EM1):111–21.
29. Kall, P., Wallace, S.W.: Stochastic programming. John Wiley & Sons, Chichester (1994)
30. Kiefer, J., Wolfowitz, J.: Stochastic estimation of the maximum of a regression function. Annals of Mathematical Statistics **23**, 462–466 (1952)
31. Kleywegt, A.J., Shapiro, A.: Stochastic Optimization: Handbook of Industrial Engineering. John Wiley, New York (2001)
32. Lee, K.H., Park, G.J.: Robust optimization considering tolerances of design variables. Computers and Structures **79**, 77–86 (2001)
33. Kuschel, N., Rackwitz, R.: A new approach for structural optimization of series systems. Applications of Statistics and Probability **2**, 987–994 (2000)

34. Le Riche R (2009) Projet OMD: Optimiser les systèmes complexes en présence d'incertitudes. Cahiers de l'ANR No. 2: calcul intensif.
35. Lee, S.H.,ChenW, : A comparative study of uncertainty propagation methods for black-box-type problems. Structural and Multidisciplinary Optimization **37**, 239–253 (2009)
36. Lee, J., Yang, S., Ruy, W.: A comparative study on reliability index and target performance based probabilistic structural design optimization. Computer & Structures **257**, 269–80 (2002)
37. Lee, S.H., Kwak, B.M.: Response surface augmented moment method for efficient reliability analysis. Structural Safety **28**, 261–272 (2006)
38. Lemaire, M., Chateauneuf, A., Mitteau, J.C.: Fiabilité des structures. Lavoisier, Paris (2005)
39. Lopez RH, Luersen M, Cursi ES (2009a) Optimization of laminated composites considering different failure criteria. Composites: Part B 40:731–740.
40. Lopez, R.H., Luersen, M., Cursi, E.S.: Optimization of hybrid laminated composites using a genetic algorithm. Journal of the Brazilian Society of Mechanical Engineering **31**, 269–278 (2009b)
41. Lopez, R.H., Cursi, E.S., Lemosse, D.: An approach for the reliability based design optimization of laminated composites. Engineering Optimization **43**, 1079–1094 (2011)
42. Lopez, R.H., Lemosse, D., Cursi, E.S., Rojas, J., El-Hami, A.: Approximating the probability density function of the optimal point of an optimization problem. Engineering Optimization **43**, 281–303 (2011)
43. Lopez, R.H.: Optimisation en présence d'incertitudes. Thèse de Doctorat, INSA de Rouen, France (2010)
44. Luersen, M.A., Riche, R.: Globalized Nelder-Mead method for engineering optimization. Computers and Structures **82**, 2251–2260 (2004)
45. Madsen, H.O., Krenk, S., Lind, N.C.: Methods of structural safety. Prentice Hall, Englewood Cliffs (1986)
46. Melchers, R.E.: Structural Reliability Analysis and Prediction. John Wiley & Sons, Chichester (1999)
47. Messac, A.: Physical programming effective optimization for computational design. AIAA Journal **34**, 149–158 (1996)
48. Messac, A., Ismail-Yahaya, A., Mattson, C.A.: The normalized normal constraint method for generating the Pareto frontier. Structural and Multidisciplinary Optimization **25**, 86–98 (2003)
49. Messac, A., Mattson, C.: Normal constraint method with guarantee of even representation of complete Pareto frontier. AIAA Journal **42**, 2101–2111 (2004)
50. Míngues, R., Castillo, E.: Reliability-based optimization in engineering using decomposition techniques and FORMS. Structural Safety **31**, 214–223 (2008)
51. Moller, B., Beer, M.: Engineering computation under uncertainty: capabilities of non-traditional models. Computers and Structures 86, 1024-1041 (2008)
52. Moore, R.E.: Interval analysis. Prentice-Hall, New York (1966)
53. Nair, V.N., Abraham, B., MacKay, J., et al.: Taguchi's parameter design: a panel discussion. Technometrics **34**, 127–161 (1992)
54. Nie, J., Ellingwood, B.R.: Directional methods for structural reliability analysis. Structural Safety **22**, 233–49 (2000)
55. Niederreiter, H., Spanier, J.: Monte Carlo and quasi-Monte Carlo methods. Springer, Berlin (2000)
56. Parkinson, A., Sorensen, C., Pourhassan, N.: A General Approach for Robust Optimal Design. ASME Journal of Mechanical Design **115**, 74–80 (1993)
57. Phadke, M.S.: Quality engineering using robust design. Prentice-Hall, New Jersey (1989)
58. Pogu, M., Souza de Cursi, J.E.: Global optimization by random perturbation of the gradient method with a fixed parameter. Journal of Global Optimization **5**, 159–180 (1994)
59. Rahman, S., Xu, H.: A univariate dimension-reduction method for multi-dimensional integration in stochastic mechanics. Probabilitic Engineering Mechanics **19**, 393–408 (2004)
60. Ritto, T., Lopez, R.H., Sampaio, R., Cursi, E.S.: Robust optimization of a flexible rotor-bearing system using the Campbell diagram. Engineering Optimization **43**, 77–96 (2011)

61. Robbins, H., Monro, S.: A stochastic approximation method. Annals of Mathematical Statistics **22**, 400–407 (1951)
62. Rubinstein, R.Y.: Simulation and the Monte Carlo method. Wiley, New York (1981)
63. Sacks, J., Welch, W., Mitchell, J., Wynn, H.: Design and analysis of computer experiments. Statistical Science **4**, 409–435 (1989)
64. Schuëller, G.I., Jensen, H.A.: Computational methods in optimization considering uncertainties - an overview. Computer Methods in Applied Mechanics and Engineering **198**, 2–13 (2009)
65. Sahinidis, N.V.: Optimization under uncertainty: state-of-the-art and opportunities. Computer & Chemical Engineering **28**, 971–983 (2004)
66. Sanchis, J., Martinez, M., Blasco, X.: Multi-objective engineering design using preferences. Engineering Optimization **40**, 253–269 (2008)
67. Seo, H.S., Kwak, B.M.: Efficient statistical tolerance analysis for general distributions using three-point information. International journal of production research **40**, 931–944 (2002)
68. Shafer, G.: A Mathematical theory of evidence. Princeton University Press, Princeton (1976)
69. Shapiro, A., Homem-de-Melo, T.: A simulation-based approach to two-stage stochastic programming with recourse. Mathematical Programming **81**, 301–325 (1998)
70. Souza de Cursi JE, Cortes MBS (1995) General Genetic algorithms and simulated annealing perturbation of the gradient method with a fixed parameter. In: Topping, B.H.V. (Ed.), Developments in Neural Networks and Evolutionary Computing for Civil and Structural, Engineering, 189–198.
71. Taguchi, G.: Quality engineering through design optimization. Krauss International Publications, New York (1986)
72. Taguchi, G.: System of experimental design. IEEE Journal **33**, 1106–1113 (1987)
73. Taguchi G (1989) Introduction to Quality Engineering. American Supplier Institute.
74. Torii, A.J., Lopez, R.H., Biondini, F.: An approach to reliability-based shape and topology optimization of truss structures. Engineering Optimization **44**, 37–53 (2011)
75. Tsui, K.L.: An overview of Taguchi method and newly developed statistical methods for robust design. IIE Transactions **24**, 44–57 (1992)
76. Tu, J., Choi, K.K., Park, Y.H.: A new study on reliability-based design optimization. J Mech Des **121**(4), 557–64 (1999)
77. Wets, R.J.B.: Stochastic programming: Handbook for Operations Research and Management Sciences. Elsevier, Amsterdam (1989)
78. Xiu, D., Karniadakis, G.E.: The Wiener-Askey polynomial chaos for stochastic differential equations. SIAM Journal on Scientific Computing **24**, 619–644 (2002)
79. Yang, R.J., Chuang, C., Gu, L., Li, G.: Experience with approximate reliability based optimization methods II: an exhaust system problem. Structural Multidisciplinary Optimization **29**, 488–497 (2005)
80. Yang, R.J., Gu, L.: Experience with approximate reliability based optimization methods. Structural and Multidisciplinary Optimization **26**, 152–159 (2004)
81. Yi, P., Cheng, G.D., Jiang, L.: A Sequential approximate programming strategy for performance measure based probabilistic structural design optimization. Structural Safety **30**, 91–109 (2008)
82. Yi, P., Cheng, G.D.: Further study on efficiency of sequential approximate programming strategy for probabilistic structural design optimization. Structural and Multidisciplinary Optimization **35**, 509–522 (2008)
83. Youn, B.D., Choi, K.K., Park, Y.H.: Hybrid analysis method for reliability-based design optimization. J Mech Des **125**, 221–32 (2003)
84. Youn, B.D., Choi, K.K.: An investigation of nonlinearity of reliability-based design optimization approaches. J Mech Des **126**, 403–11 (2004b)
85. Youn, B.D., Choi, K.K.: Selecting Probabilistic Approaches for Reliability Based Design Optimization. AIAA Journal **124**, 131–42 (2004a)
86. Zadeh, L.A.: Fuzzy sets. Information and control 8, 338-353 (1965)
87. Zhang, Q., Li, H.: MOEA/D: a multiobjective evolutionary algorithm based on decomposition. IEEE Transactions on evolutionary computation **11**, 712–731 (2007)
88. Zimmermann, H.J.: Fuzzy Set Theory and its Applications. Springer, Berlin (1992)

Printed by Publishers' Graphics LLC
DBT131023.15.14.30